ULTRAVIOLET

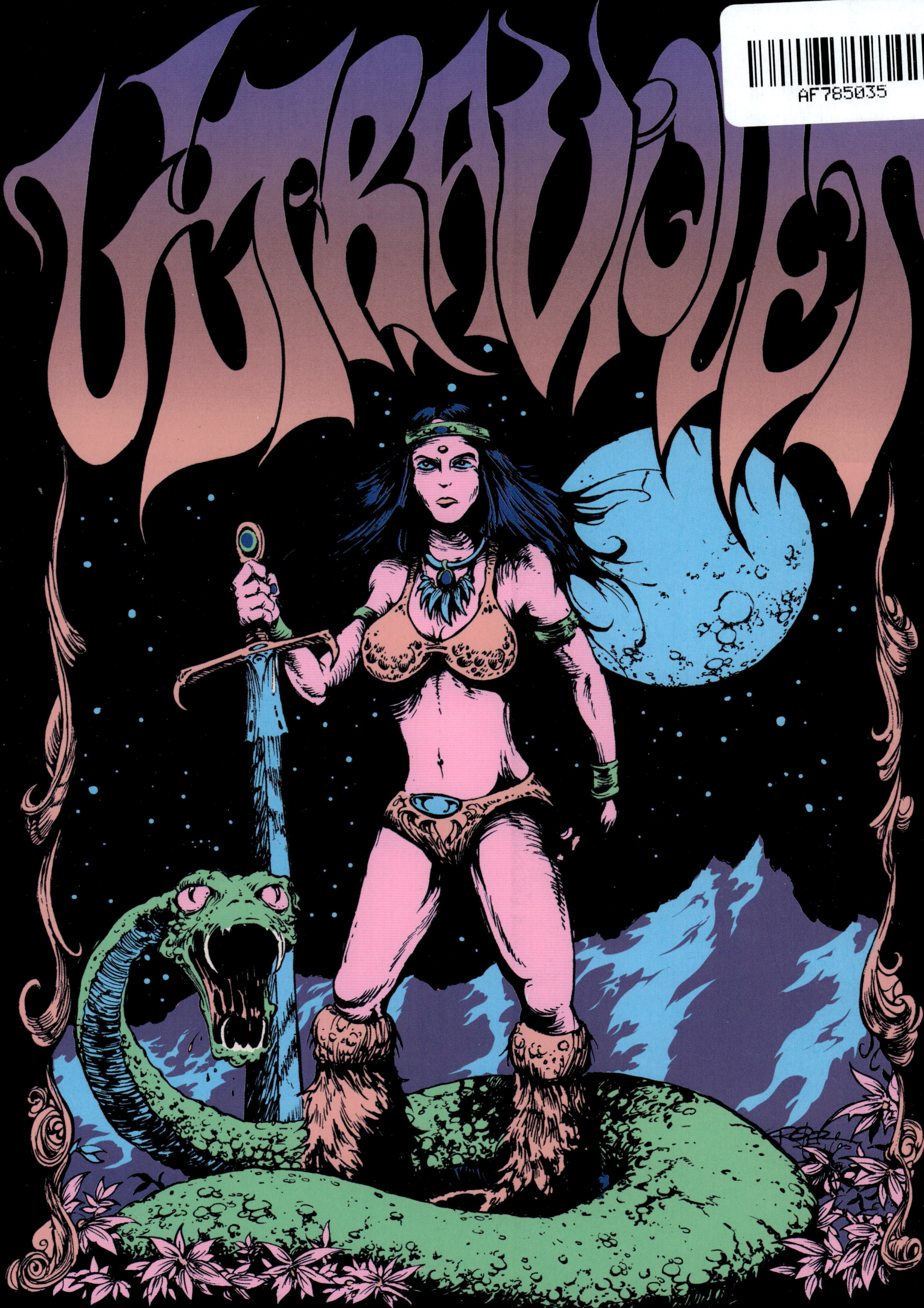

69 BLACKLIGHT POSTERS FROM THE AQUARIAN AGE AND BEYOND

DAN DONAHUE

ABRAMS IMAGE • NEW YORK

Editor: David Cashion
Art Director: Dan Donahue at www.controlgroop.com
Interior layout designer: Liam Flanagan
Production Manager: Jacqueline Poirier

Cover illustration by Dan Donahue
Title page illustration by Arik Roper at www.arikroper.com
Foreword illustration by Jeff Sims

Cataloging-in-Publication data has been applied for and is available from the Library of Congress.

ISBN 978-0-8109-7999-4

Printed and bound in China
10 9 8 7 6 5 4 3 2 1

ABRAMS
THE ART OF BOOKS SINCE 1949
115 West 18th Street
New York, NY 10011
www.abramsbooks.com

Opposite: *Atomic Exploratory System*, 1968, Clard Svensen, The Third Eye, NY, NY

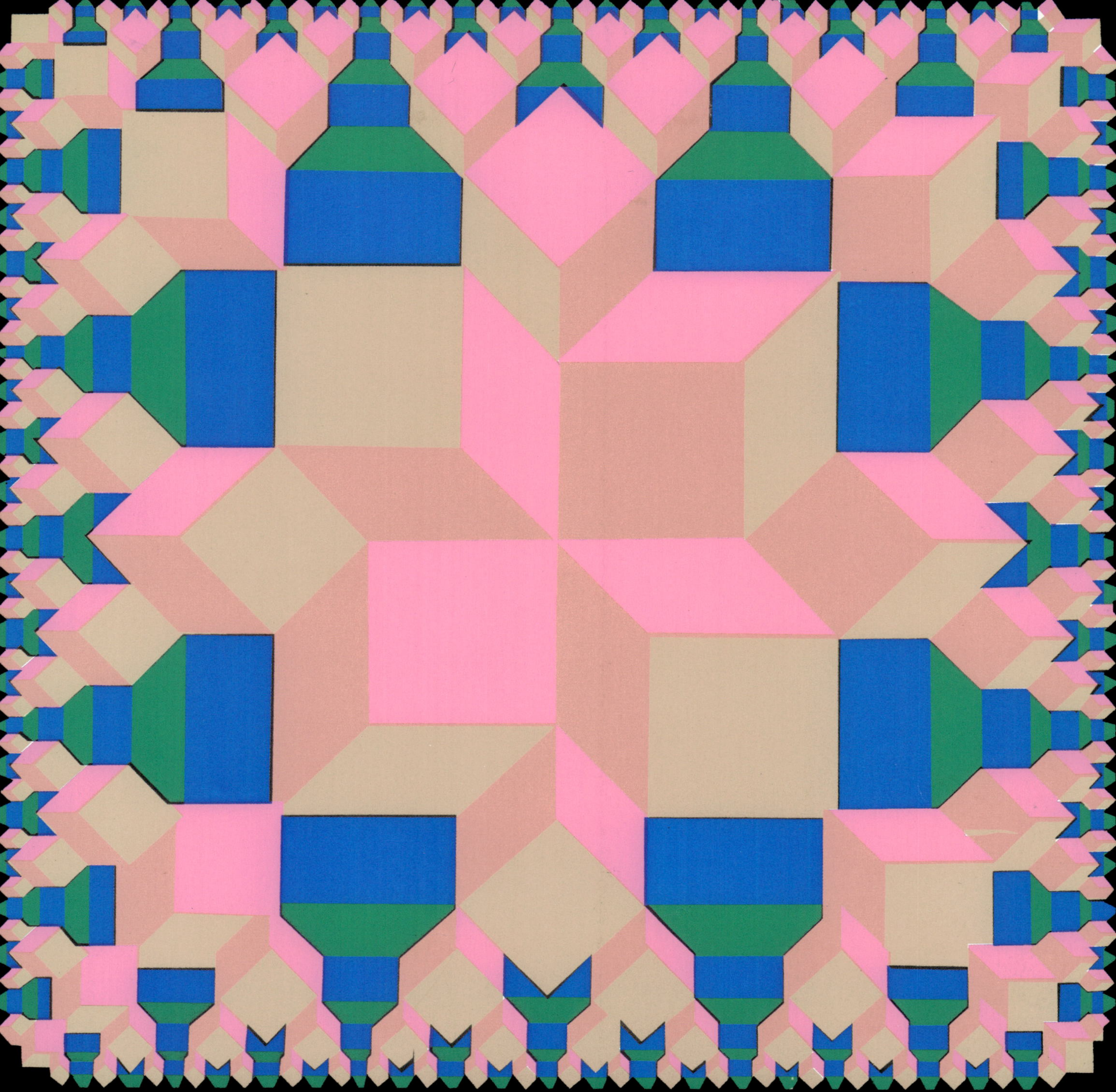

FUZZY WUZZY WUZ THE PINSTRIPE CLOAKED **WIZARD** THAT WATCHED PROUDLY O'ER OUR GALLANT PRACTICE SPACEY SPACE AT SUCH PERILOUS HEIGHTS. WHAT IMAGINED CRAG DOTH HE SCALE AT DAWN, STAFF IN HAND? BEFORE THE ROILING SEAS, IMPERFECT TEETH ALL **AGLOW** AND DRY-WALL DUST-COVERED HEATHER GREY SWEATS, HE DOESN'T SWEAT IT; HIS IS TO ACQUIESCE ALL THAT DO **FLOURESCE.** TWO PUSH PINS AND A STRIP OF SCOTCH TAPE AND BEFORE OUR HAIRY NOSES CAN DUCK AND HIDE, ***THE SPELL IS CALLED UPON AND CAST***: THE **LIGHT** THAT IS **BLACK** AND ALL IT DOES **REVEAL.** FLICKER ON THE **ULTRA VIOLET**; LINES GET EXCITED FOR A BIT OF THE OLD IN-OUT IN-OUT, WHILST **COLORS** PUMP AND JITTER BEFORE NERVOUS **OCULAR LUST.** L'ORANGES, PINKS, AND ICE CUBE, BUT THE PAPER AND INK ON THE WALL IS THE LAST-GLEAMING STAR, GLIB WITH THE THOUGHT THAT IT'S NOT NICE TO FOOL MOTHER NATURE.

ALAS, HERE ENTERS THE ***FATAL CIRCUMSTANCE***:

"WHAT?"

"THE FATAL CIRCUMSTANCE(S)."

"I DO NOT UNDERSTAND."

"NOT ALL MEN CAN BE HEROES."

MEIN MILD FERVENT HOT DOG **BRETHREN** WITH HIS BAG OF MONEY WITH ITS ***"DOLLAR SIGN"*** MARKED PLAINLY, BLUE BLACK INK ON DIRTY CANVAS, WITH THE LUST IN HIS MEATY EYE. WINDOW SHOPPING ON MULBERRY ST, LITTLE ITALY.

"I'M ONLY LOOKING."

"IT'S AN 'S' WITH TWO LINES THROUGH IT!"

WHO CAN MAKE THE ***DAY GLOW LONGER***? COMMAND REVERENCE THAT DWARFS THE JOINT-SMOKING GREEN ALIEN IN THE STOCKING CAP? SPRINKLE IT WITH DEW? NO, THIS IS FAR ***BEYOND*** 311, GREATEST HITS OR NOT. THIS IS WORTH TAKING THOUGHT OF:

DAN CAN.

A ***WISE*** AND ***WEIRD*** SOMEONE ONCE ALMOST SAID:

"FIND A MAN WHOSE [BLACKLIGHT POSTERS] PAINT YOU A LIKENESS, YOU HAVE FOUND A MAN WORTH SOMETHING."

MGMT

TURNED ON

THE IMAGE GETS ELECTRIC

By the summer of 1967, there was a great wave of change n the United States. More than a hundred thousand hopefuls had made a pilgrimage to the Haight-Ashbury neighborhood of San Francisco to lose themselves and find nirvana in an oasis of experimental music, sexuality, politics, and drugs. It was a very public display of a scene that would soon be playing out all over the country. But before the so-called Summer of Love could take place, a seed for the season needed to be planted. Psychedelic sage and psychologist Timothy Leary gave the growing movement a phrase to follow: "Turn on, tune in, drop out." He delivered this permissive mission statement in September of 1966 at the infamous and pivotal event known as "The Human Be-In." Turning on (i.e., toking, smoking, popping) was the first step in the transformation of the culture to a counterculture. Bob Dylan turned the Beatles on to marijuana and turned on himself by playing electric in the face of folk. Andy Warhol turned on and turned on the art world, transforming American icons like soup cans and Brillo boxes into day-glo art at his decadent creation space, the Factory. The squeaky clean and streamlined ideals of the previous decade were dying. The psychedelic movement had taken hold as people were beginning to see things in a new light, and for many, it was under a blacklight. Everyone from doctors to dropouts was seeking illumination from everything under the sun, but more often than not, it was in the darkness that they were finding it. The blacklight poster had become the premier freak flag to hang at the front door to a new consciousness.

The blacklight version of turning on is flipping the electric switch and waiting for the unmistakable hazy glow. Ultraviolet light from a bulb was the key to transforming the poster's visions and exciting the eyes. The power of the blacklight is as sci-fi as it sounds, using ultraviolet light with a wavelength that sits between visible light and the X-ray spectrum to expand the visual range to the naked eye. Used in combination with outrageous illustrations, the light created a glorious glow and spectacular optical odysseys.

The origins of the blacklight revolution, however, can be traced to the early twentieth century. An early incarnation of the modern blacklight bulb was possibly something known as "Wood's glass"—originally used in World War I for ultraviolet night communications—which filtered visual light and revealed the invisible radiation as the communication source. Not just for spies or wartime espionage, though, Wood's glass was also useful in the world of medicine. The bulb is able to locate the presence of certain fungi that cause skin infections, such as the dermatophyic strains that emit a yellow or

Then, in 1934, experimental teenagers found a new use for this power. Eighteen-year-old Joe Switzer was playing with illumination in his father's drugstore when a combination of ultraviolet light and some chemical compounds resulted in a glow that captivated him and his brother, Bob. Curious, he combined the compounds with shellac, and a primitive blacklight paint was created. The brothers debuted their creation at a magic show where painted objects were made "magical" and glowed under their bulb. Ecstatic by the early excitement their creation had caused, these enterprisers started "the Switzer Brothers Ultra Violet Laboratories Company Inc." in their mother's Berkeley kitchen. They created a mini-empire that saw their developments used for the military, but also benefit lighter enterprises from swimwear to Hula-hoops to everyday packaging.

Thirty-some years later, their happy accident had spawned posters emblazoned with pop-culture icons, geometric patterns, antiwar statements, and acid-trip art, providing the backdrop to radical rallies, existential experimentation, and free-love freak-outs nationwide.

In the 1960s, LSD, or Lysergic acid diethyl-amide, was the principal agent of change: Its mind-expanding properties had literally changed the way people saw the world and the way they expressed their visions. It caused heightened sensations, visual vibrations, and hallucinations. Many even felt in touch with religious entities and had visions. The blacklight poster was actually a medium capable of mimicking the effects of the new wonder drug. With the ability to glow and vibrate under ultraviolet light, the posters could simulate the sensations and visual distortions one experienced during an acid trip. Allen Cohen, publisher of the *San Francisco Oracle* and one of the blacklight medium's early champions took full advantage of this and published covers of his weekly alternative newspaper with wildly psychedelic blacklight masterpieces. Some issues included massive mandalas (the cosmically concentric centerpieces used in rituals, meditations, and trances) that were printed as centerfolds with instructions on how to "meditate with blacklight mandala and get high without getting freaked by the heat!" These early advocates of visual stimulation pushed posters on the people, exhorting them to expand their consciousness without substances, claiming the blacklight as a "fantastic cosmic brain turn-on—and it's legal."

This invitation to meditation asked followers to disrobe and sit in darkness with only a blacklight for illumination. "You are taking yourself on a trip beyond the game-world of prestige, status, achievement, suc yourself off and get high with yourself. You are going to return to the Source, to the Essence, to the Atman, the Godhead within, Source of all creation, all powers, all energies, the intuitive center." The allure of getting "high, high, very high" by using a technique "similar to most yoga, but much easier" was further promoted as a way "you, too, can turn on to meditation, just like the Beatles, Rolling Stones, and all those other righteous cats." This idealistic and noble goal was amended at the end of the article to include the hazy hippies on Haight: "The technique just described can be useful as a means of centering yourself on trips with psychedelics such as LSD, mescaline, psilocybin, peyote, or grass." The *Oracle* was such an advocate of the blacklight movement and the psychedelic art movement in general that they formed the Jewel-Mirror Radiation Company, their blacklight poster company, which solicited contributions from readers, offering lifetime subscriptions as a reward.

The advertisements inside the *Oracle* were a virtual guide to head shops across America. These stores were the original galleries of blacklight posters and underground art. With names like Head East, Trip Shop, Mind Mart, and the Electric Lotus, their intent was as thinly veiled as the subjects of early blacklight, head-popping posters. They were also the perfect place to get your blacklight bulb or tube, or maybe a beer can lamp to house it. Head shops often evolved from old or failing gift stores, candle shops, or poster shops, and became the bazaar of the bizarre, often inviting youth to seek their first psychedelic experiences. The positioning of the blacklight poster—often in back rooms and dark corners—provided a safe place to view alternative, taboo, and even perverse subject matter.

At the head shops, you could expand your experience with adult novelties and pot paraphernalia hanging on the walls, or, on the counters, you could find copies of the underground press from east and west, including San Francisco's *Oracle* and New York City's the *East Village Other*. The latter paper, or *EVO* as it became called, was decorated with psychedelic covers like its West Coast counterpart and was an early publisher of underground comic artists such as Spain Rodriguez, Art Spiegelman, and wimmen's comix artists Trina Robbins and Kim Deitch. *EVO* published it's own comic, *Gothic Blimp Works*, "the first Sunday underground comic paper," first edited by the legendary Vaughn Bode, whose style set the tone for countless blacklight posters as well as creating a style siphoned by seventies graffiti artists. The *EVO* also featured the work of Robert Crumb, who often cited the LSD experience as having affected his illustration style and whose *Keep on Truckin'* image was one of the most iconic blacklight posters of the entire era. The subjects of these comics being metaphysical wizards, acid-heads, and antiheroes battling on campuses, outer space, or fantasy landscapes, were a primary influence on the out-there art of the early blacklight posters.

The head shops were the first to cash in on the new poster craze, but not the last. The combination of a newly affluent middle class, the climate of experimentation, and an exploding pop-art movement provided the soil for a growing cottage industry. The phenomenon prompted *Life* magazine to devote their September 1, 1967, cover to a story called "The Big Poster Hang Up: Walls and Walls of Expendable Art."

This so-called expendable art galvanized disenfranchised youth and turned basements and bedrooms all over America into makeshift opium dens or hippie hangouts where many joints were rolled ... acid

rock albums were spun while bathing in the blacklight. Poster presses spread throughout the West Coast and eventually across the country to take advantage of the growing demand. The counterculture was becoming dominant.

Another beneficiary of the blacklight movement was the music scene. Concerts became life-changing "experiences" augmented by film loops, day-glo body painting and oil lamp light shows performed by such artists as Headlights, the Single Wing Turquoise Bird, and the Brotherhood of Lights, who created light shows at San Francisco's Winterland and Fillmore West for the Who, Santana, the Grateful Dead, Led Zeppelin, and Jimi Hendrix, among others. Blacklight and body paint were frequently elements of the festivities, giving free love a fluorescent shine. Bodies would twirl, making swirling trails for the heavily dosed to enjoy.

Posters for these shows were highly and wildly designed, frequently with a heavy dose of art nouveau style, a look that had been inspired by a then-recent retrospective on the era held in Berkeley, California. Pioneers in the new music poster medium included Stanley Mouse, who was raised on custom car culture in the 1950s and had mastered a menagerie of monster characters and outrageous hand-lettering, as well as Alton Kelley, an art-school dropout who worked on and raced motorcycles. The two of them formed Mouse Studios and expanded and exploded the possibilities of the rock-show poster. They mixed wild type styles and collage, and adapted advertising icons such as the Zig-Zag man for their own devices. (The Zig-Zag man, whose smoking head became the unofficial icon of the blacklight art form, found his way into countless designs, whether as the subject or the secret signal that revealed you were going to get high on this piece. Victor Moscoso was one of the only classically trained graphic artists in the psychedelic scene. He was trained at Cooper Union and at Yale where he studied with renowned color artist Josef Albers. He was one of the most experimental of all the artists with his use of "hot" color to electrify the surface and create what he called the "vibrating border." His illustration style could only be called mind-bending, as his detailed style literally twists its characters and the reality in which they sit until the viewer has no sense of ceiling or ground, up or down. He also developed a lettering style which became synonymous with psychedelic posters, using chunky heavily serifed characters mimicking saloon characters and giving a way-out, wild west feeling to his fonts.

Many poster artists did album covers for recording artists like the Grateful Dead, Janis Joplin, and the Velvet Underground. Rick Griffin, a premiere poster artist who embodied the California way of life and who dropped acid with the Merry Pranksters, created comics, magazine illustration, as well as album artwork. He did covers for the *Oracle* and for *Rolling Stone*, whose original logo was his handiwork. Rick embodied the California counterculture, and in 1967 he formed the Berkeley Bonaparte Society to promote and produce psychedelic and posters. His lettering and highly stylized compositions became the archetype for the psychedelic poster and his illustrations shone beautifully in blacklight. Far from the West Coast was Stanislaw Zagorski, the Polish poster designer and illustrator whose work on famous film posters and with graphic gurus Push Pin Studios was internationally renowned and revered. For Velvet Underground's swan song opus, *Loaded*, he created the iconic image of a violet smoke swirl escaping the

New York City subway, which begs to be seen in UV light. His most famous blacklight poster, *Fly Carefully*, portrays a fatalistic vision of the space race craze: Its Icarus spaceman with rainbow wings appears to have peaked in his flight to a stop sign sun, while exploding and expanding traffic signs radiate around him. Fading into space, the words "Do Not Enter" hang; a perfect message for a zoned-out stoner's door.

One of the hometown heroes of the Haight-Ashbury scene was Funky Features, a printing company whose images ranged from psychedelic collages and op-art meditations to intensely detailed, introspective illustrations. They ran their business out of a large Edwardian home and delivered their blacklight posters in a mildly disguised postal van bought at auction from the government and used to confuse it shortly after. Cops didn't notice the truck, as the only change was the addition on both sides of "U.S. FREAK 6969" in a government issue–style font. From 1966 to 1969, the company was owned by the three friends and artists "Funky Paul" Olsen, "Funky Sam" Ridge, and "Funky Jack" Leahy. Paul Olsen was an artist whose first LSD trip made him quit his job at a lithography studio and create high jinx on the Haight by painting hydrants in wild rainbow colors with a group called the Psychedelic Raiders. He was the one who gave the three their names and posed pant-less in the company photo, his wild side showing in everything from their business cards to their blacklight posters.

Some artists veered away from the letter-heavy designs of the rock show poster or album covers. Harry Hambly of Hambly Studios was one of those artists and was also in attendance at the Human Be-In and Leary's words were not wasted on him. Harry chose to turn on with the pulsating power of these new colors and their combinations. Arming his assistants with dental robes obtained from his father's practice, Harry and company set out to create wild color combinations. Hand-pulled screens combined with over-excitement for the new world of hues they were creating often caused them to be covered in colors. He would often work his assistants into the art, having them portray fringe elements in city scenes. First they would test color combinations indoors under blacklight, and then out in the sunlight where these colorful characters were often the subjects of stares from the employees of the Memorex corporation, located across the street. Combing crazed, day-glo colors and more peaceful pastels, the images they created often had disorienting spatial relations, making the subjects almost dance with the background. Hambly Studios applied its wild, wide-eyed color creations to ancient etchings and Harry's own cartoon creations.

The end of the sixties left once hopeful hippies and resistant radicals disillusioned and disengaged. Madison Avenue had adopted every trick in the psychedelic handbook, and artists like Peter Max became household names, designing clocks for General Electric and jeans for Wrangler. The 1970s saw a marked departure from the medium having a message to it just being a consumable product for anyone, whether they turned on or not. Pop culture propaganda and mass merchandising had replaced the protest of a decade before. Although no less trippy, the posters saw a turn toward the fantastic and the famous. The Third Eye company in New York, known for its blacklight optical-art posters and sculptures, was producing posters and greeting cards with Marvel super heroes saving the world in screaming blacklight inks. The pop-art dream had come full circle, with comic icons becoming pop products in this capitalist escapist atmosphere. The music had shifted from be-ins and festivals to stadium rock, and the big bands were now big business. No longer a bootleg article, the blacklight poster had become an attractive commodity for mass consumption, and the acts often became larger-than-life creatures and characters whose wild styles were almost calculated to be immortalized in inks. Savvy bands like KISS produced a world of face-painted products including the inevitable blacklight poster. These extremely eccentric, larger-than-life-characters fit fashionably with the spacing out of the seventies.

Science-fiction stylings were taking over with barbarians and beauties prowling fantasy landscapes and weird warriors riding massive, psychedelic cycles. By the late seventies, *Conan* book covers, *Vampirella*, and *Heavy Metal* magazine had made the blend of sex and sorcery an accepted art, while pinball machines adorned with buxom beauties and way-out wizardry were the arcade standard. Video games followed suit and featured side panel art with laserlike letters and trippy spacescapes. A new glow possessed the youth of the nation. With such fantastic fodder for blacklight experiences, arcades became dark dungeons with fluorescent facades and blacklight wallpaper.

In its young history, the intense imagery found in blacklight art has been burned into the teenage psyche and continues on as outsider art still inspiring the modern design world. From the energetic and elastic letters of early graffiti to the day-glo palette of iconoclastic designers such as Stephen Sprouse in the eighties and nineties and, more recently, Marc Jacobs, its potent power still shines. The rave revolution stole its day-glo smiles straight from the walls of the seventies, the last face of a dying decade in the midst of disco. The eighties also saw electric color combinations continue to contort in the world of skateboarding and deck graphics. Often optic and grotesque, this aggro outsider art affected the kids and the culture by making day-glo logos and screaming skulls commonplace.

This energy has not been displaced in the twenty-first century and vibrant visual-artist collectives such as Dearraindrop and Paper Rad have adapted the spectrum of the medium and kaleidoscopically collaged a culture of day-glo excesses and cosmic comics. These new voices and visions signal that it is time to let your blacklights shine.

Shine on.

—Dan Donahue

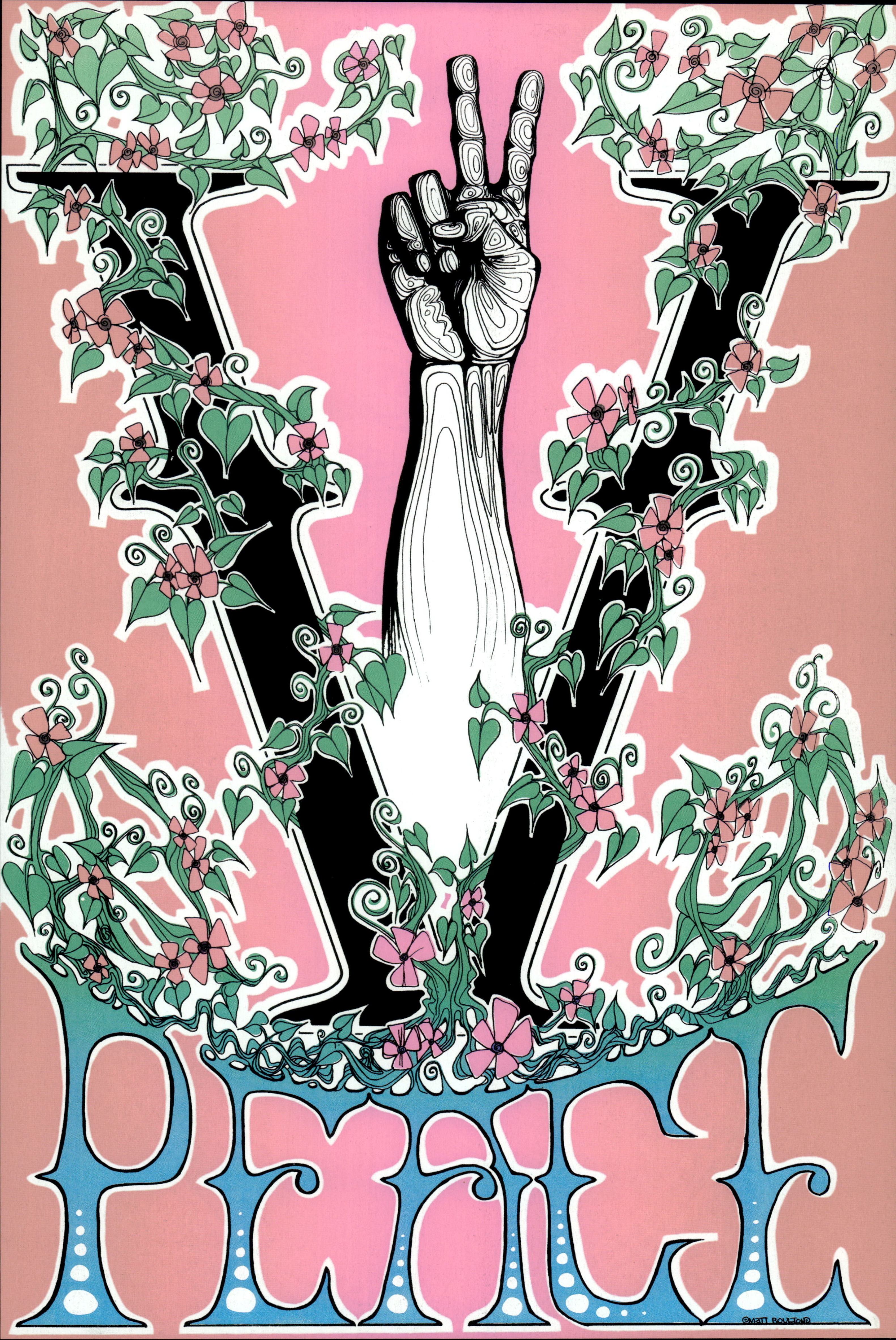
PEACE
©MATT BOULTON

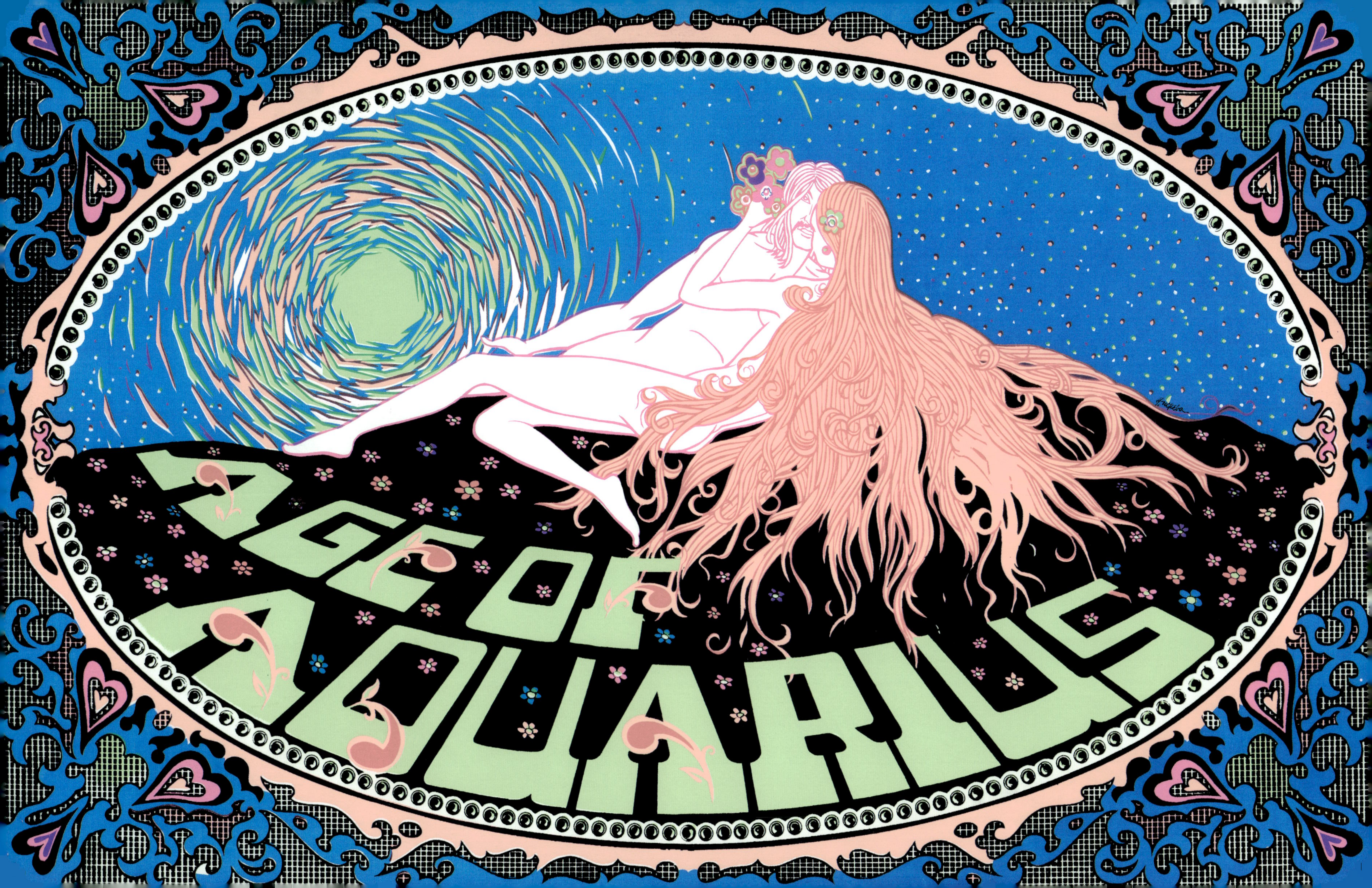
AGE OF AQUARIUS

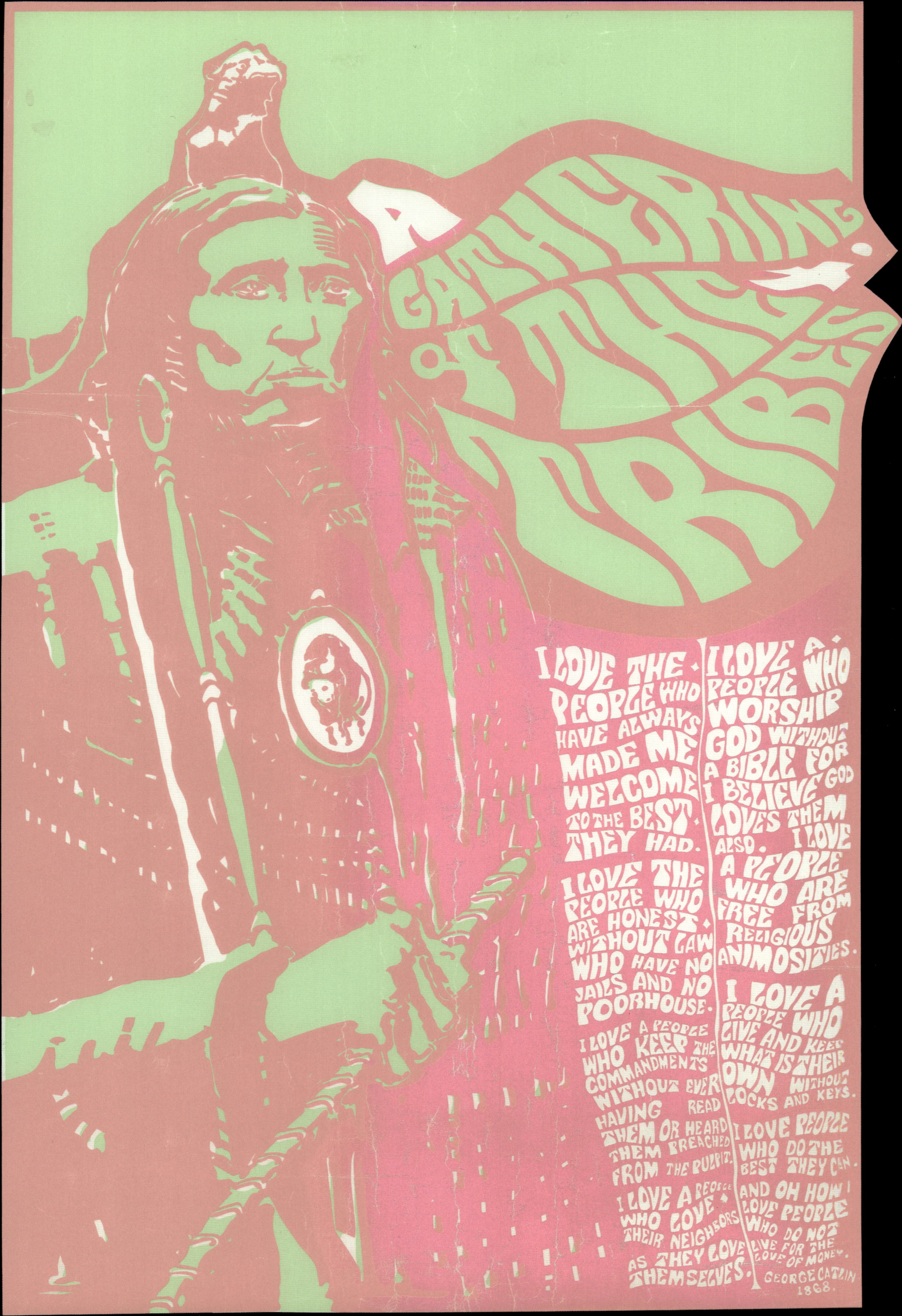
A GATHERING OF THE TRIBES
I LOVE THE PEOPLE WHO HAVE ALWAYS MADE ME WELCOME TO THE BEST THEY HAD.
I LOVE THE PEOPLE WHO ARE HONEST WITHOUT LAW WHO HAVE NO JAILS AND NO POORHOUSE.
I LOVE A PEOPLE WHO KEEP THE COMMANDMENTS WITHOUT EVER HAVING READ THEM OR HEARD THEM PREACHED FROM THE PULPIT.
I LOVE A PEOPLE WHO LOVE THEIR NEIGHBORS AS THEY LOVE THEMSELVES.
I LOVE A PEOPLE WHO WORSHIP GOD WITHOUT A BIBLE FOR I BELIEVE GOD LOVES THEM ALSO. I LOVE A PEOPLE WHO ARE FREE FROM RELIGIOUS ANIMOSITIES.
I LOVE A PEOPLE WHO GIVE AND KEEP WHAT IS THEIR OWN WITHOUT LOCKS AND KEYS.
I LOVE PEOPLE WHO DO THE BEST THEY CAN.
AND OH HOW I LOVE PEOPLE WHO DO NOT LIVE FOR THE LOVE OF MONEY.
GEORGE CATLIN 1868.

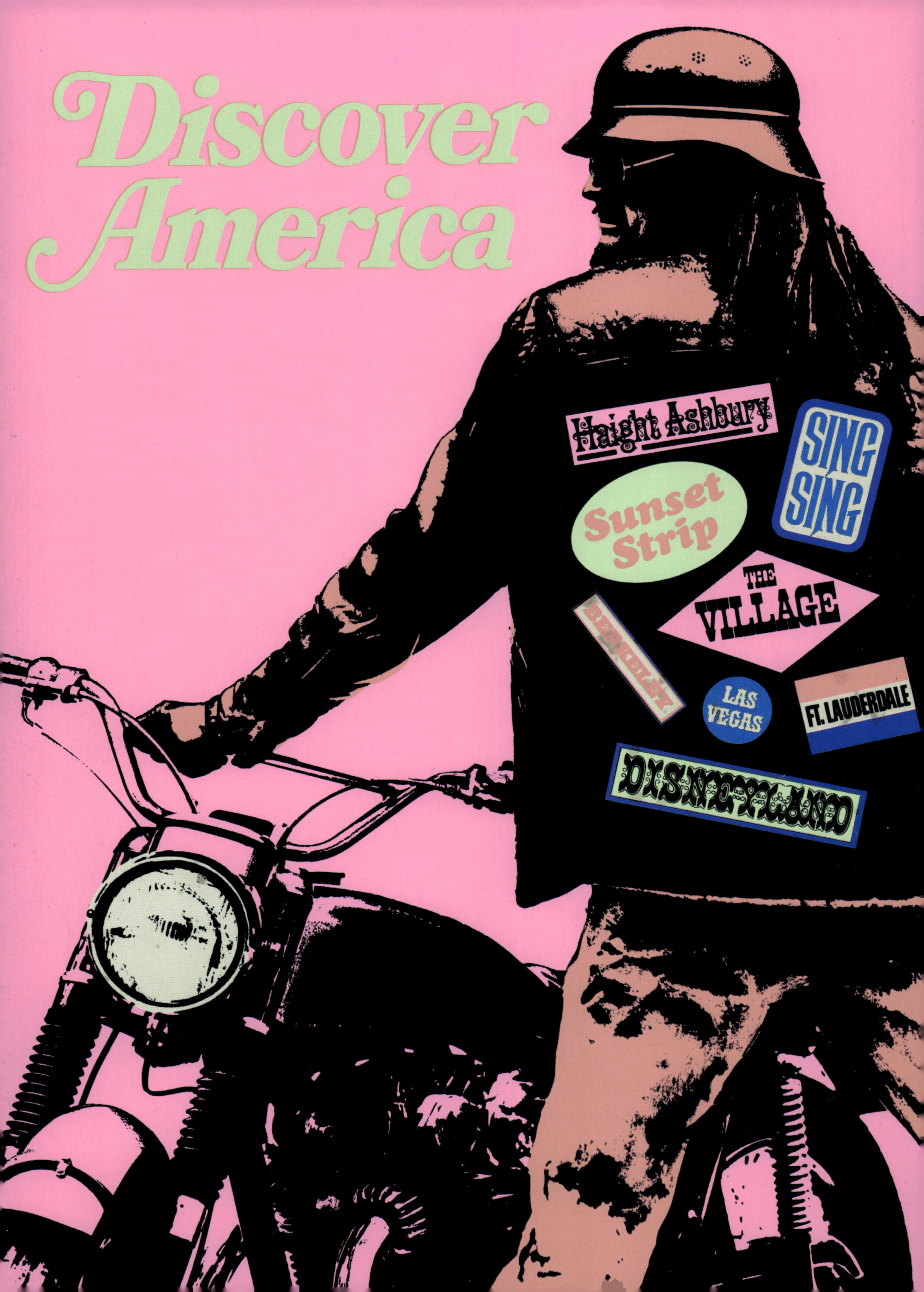
Discover America
Haight Ashbury
SING SING
Sunset Strip
THE VILLAGE
LAS VEGAS
FT. LAUDERDALE
DISNEYLAND

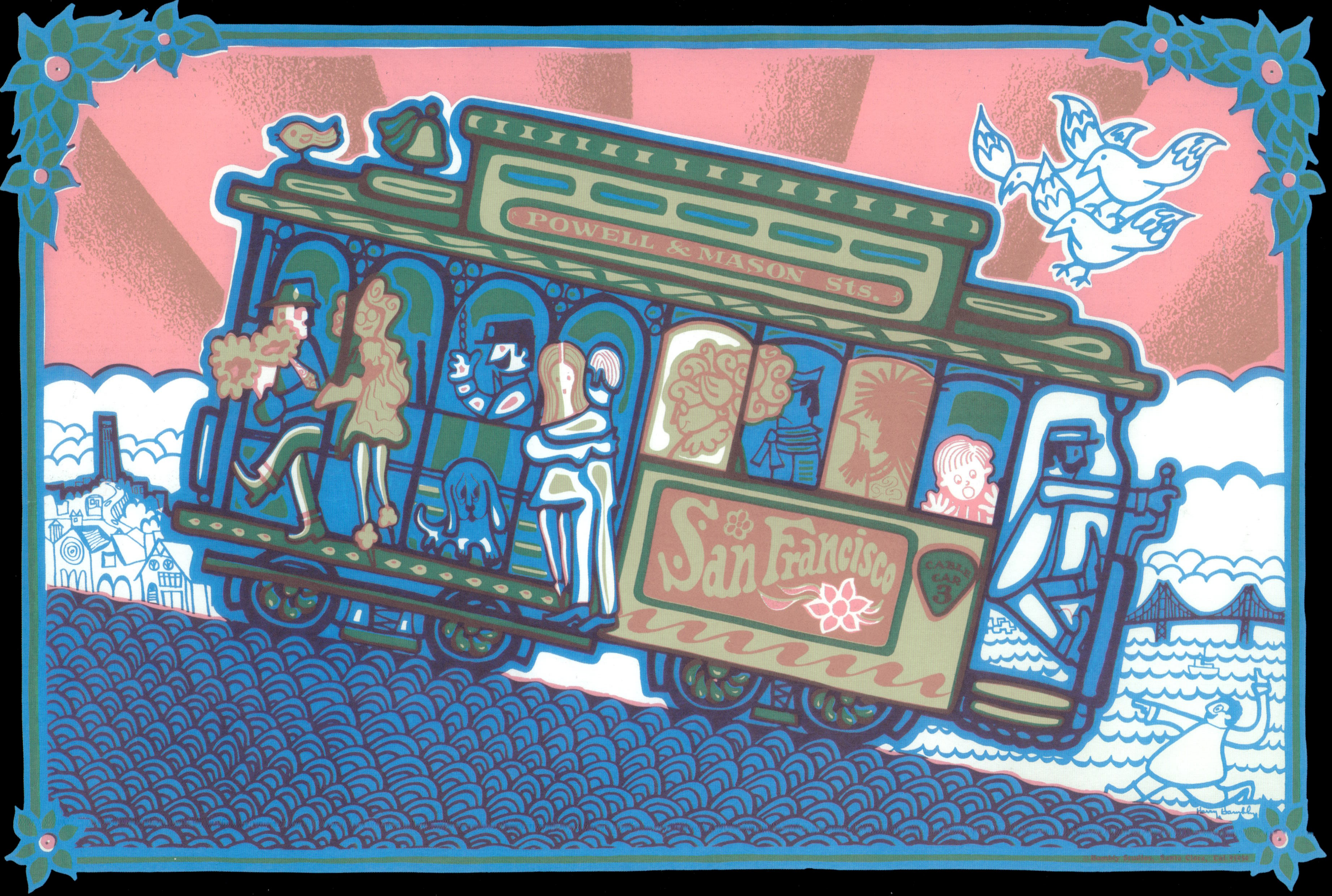
POWELL & MASON Sts.
San Francisco
CABLE CAR 3

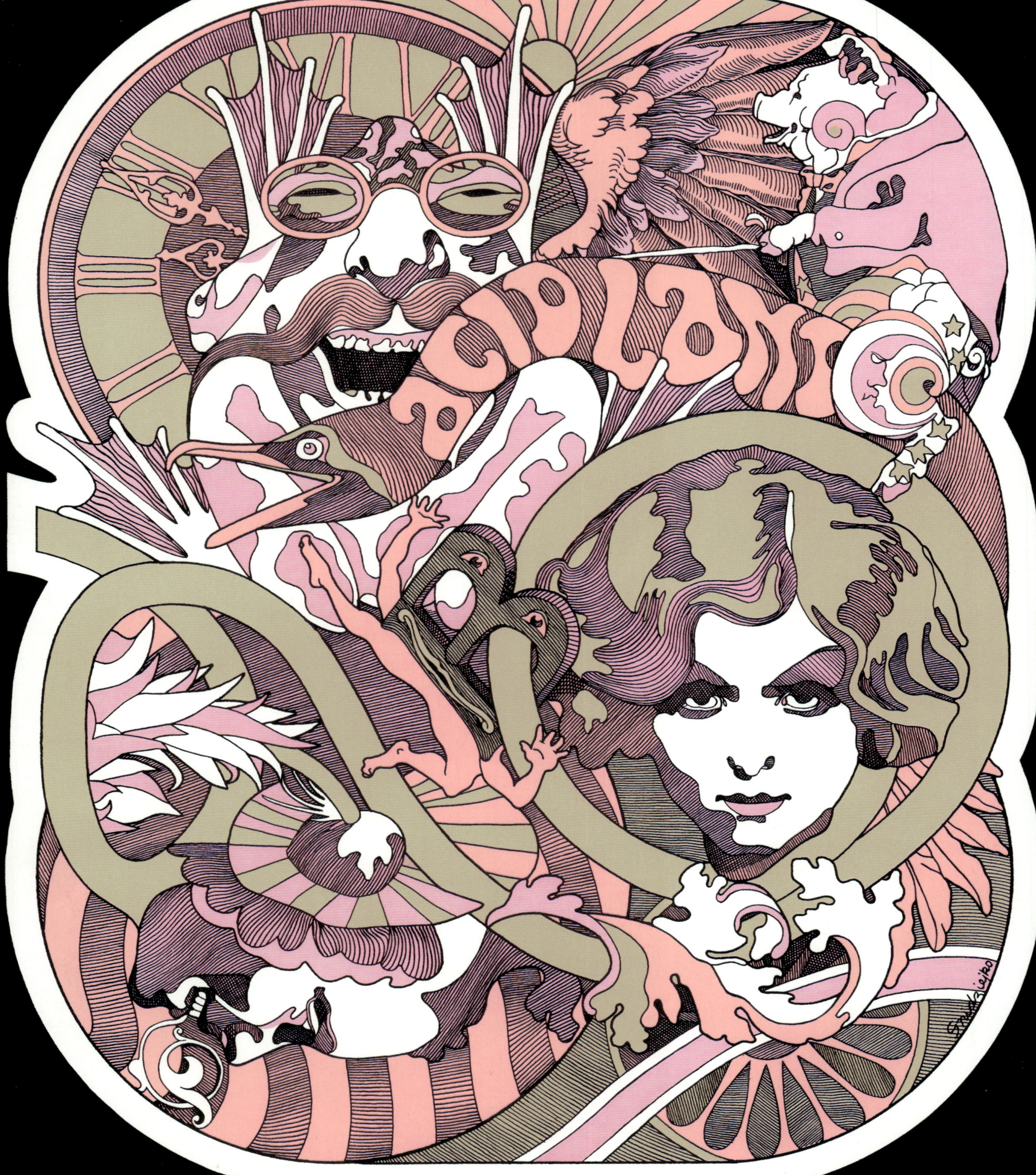

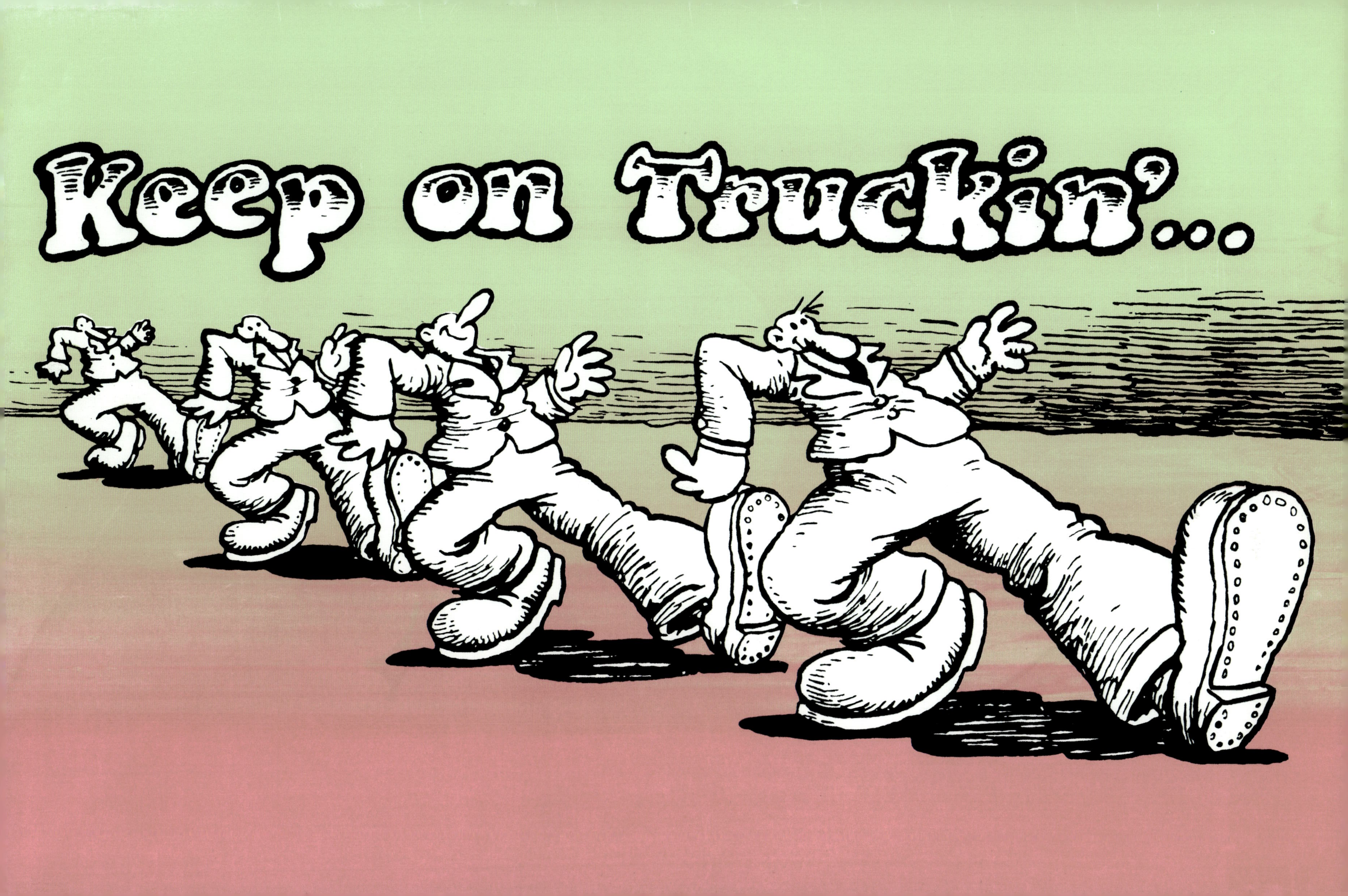
Keep on Truckin'...

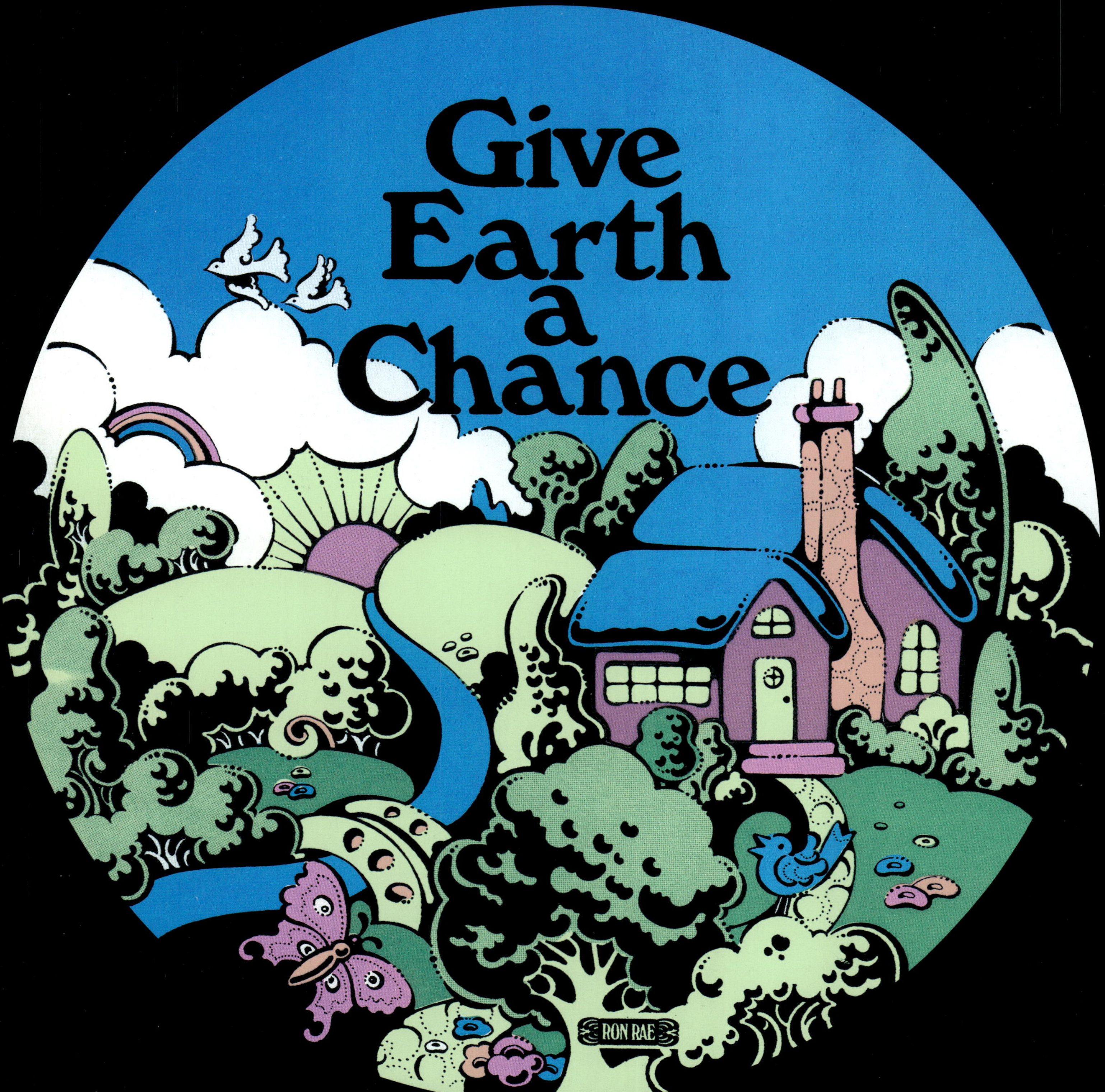
Give
Earth
a
Chance
RON RAE

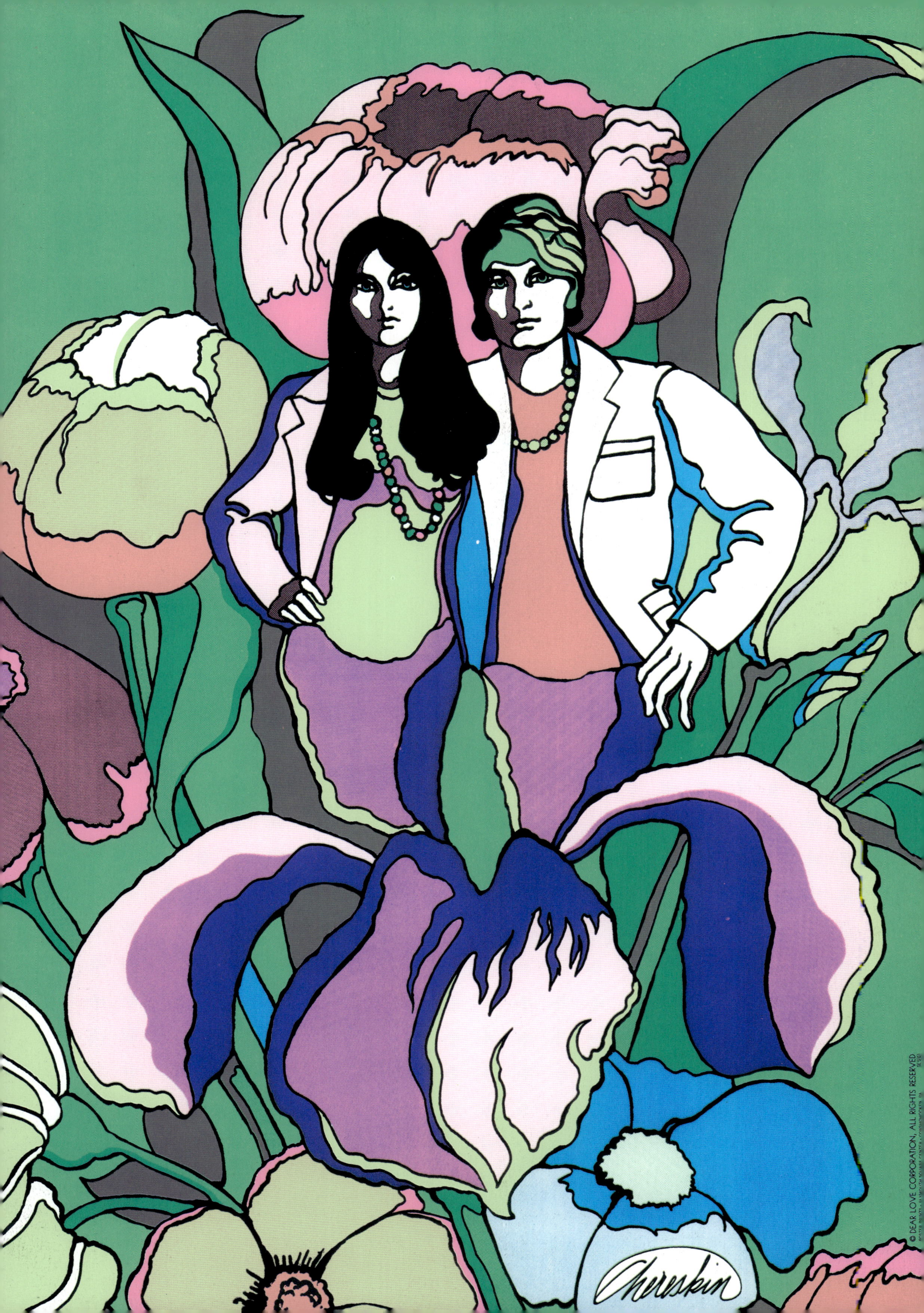

© DEAR LOVE CORPORATION, ALL RIGHTS RESERVED
Chereskin

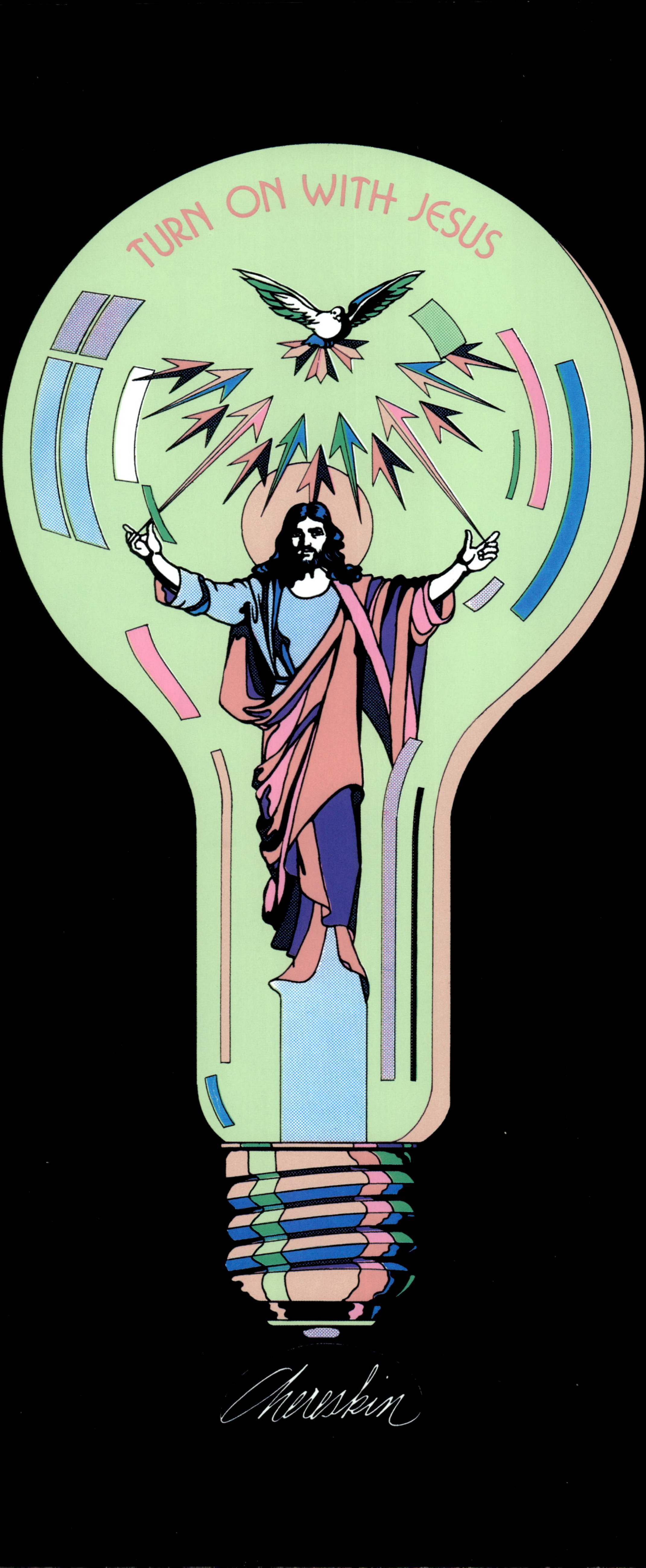
TURN ON WITH JESUS
Chereskin

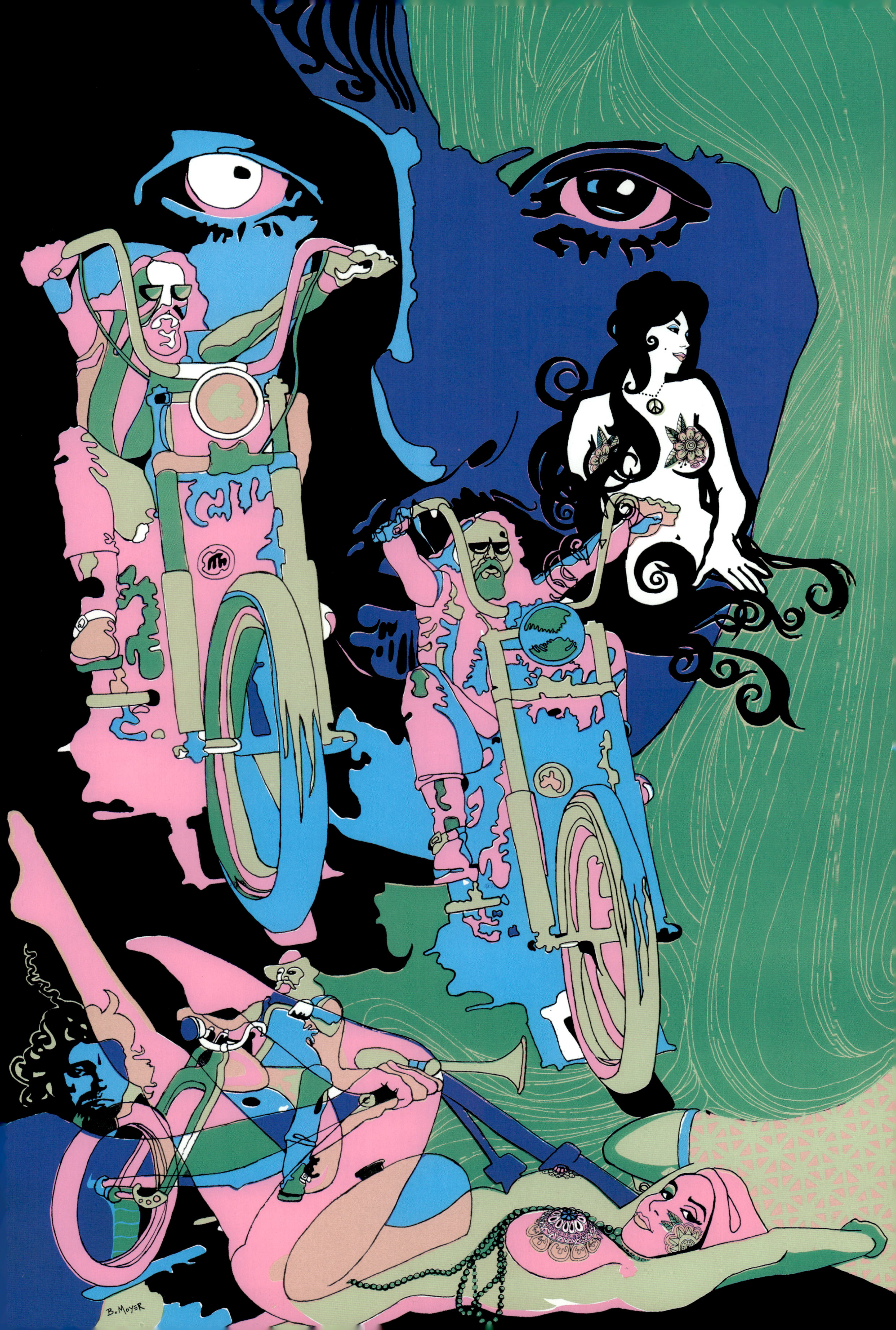
B. MOYER

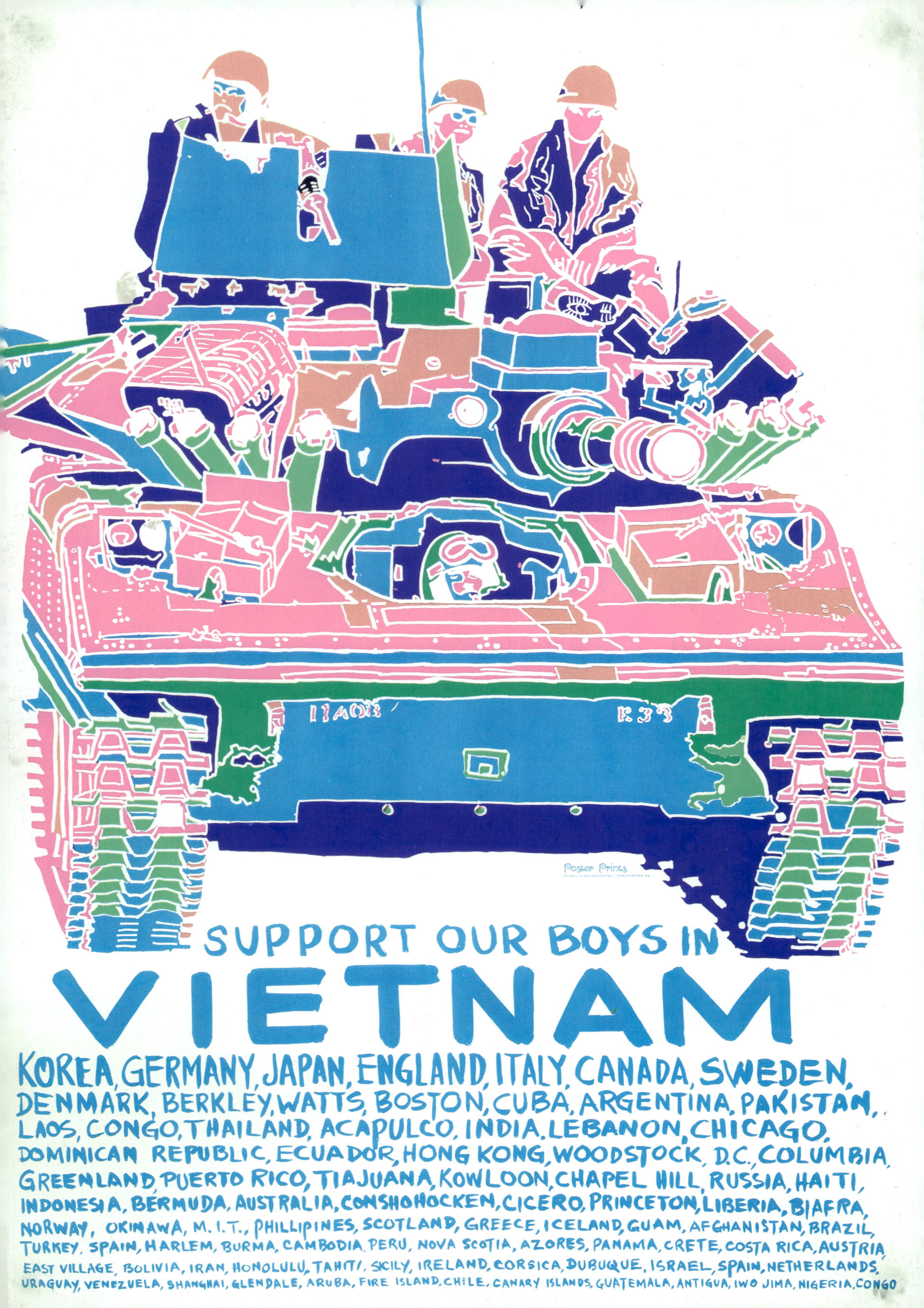
11AO8
E 3 3
Poster Prints
SUPPORT OUR BOYS IN
VIETNAM
KOREA, GERMANY, JAPAN, ENGLAND, ITALY, CANADA, SWEDEN,
DENMARK, BERKLEY, WATTS, BOSTON, CUBA, ARGENTINA, PAKISTAN,
LAOS, CONGO, THAILAND, ACAPULCO, INDIA, LEBANON, CHICAGO,
DOMINICAN REPUBLIC, ECUADOR, HONG KONG, WOODSTOCK, D.C., COLUMBIA,
GREENLAND, PUERTO RICO, TIAJUANA, KOWLOON, CHAPEL HILL, RUSSIA, HAITI,
INDONESIA, BERMUDA, AUSTRALIA, CONSHOHOCKEN, CICERO, PRINCETON, LIBERIA, BIAFRA,
NORWAY, OKINAWA, M.I.T., PHILLIPINES, SCOTLAND, GREECE, ICELAND, GUAM, AFGHANISTAN, BRAZIL,
TURKEY, SPAIN, HARLEM, BURMA, CAMBODIA, PERU, NOVA SCOTIA, AZORES, PANAMA, CRETE, COSTA RICA, AUSTRIA,
EAST VILLAGE, BOLIVIA, IRAN, HONOLULU, TAHITI, SICILY, IRELAND, CORSICA, DUBUQUE, ISRAEL, SPAIN, NETHERLANDS,
URAGUAY, VENEZUELA, SHANGHAI, GLENDALE, ARUBA, FIRE ISLAND, CHILE, CANARY ISLANDS, GUATEMALA, ANTIGUA, IWO JIMA, NIGERIA, CONGO

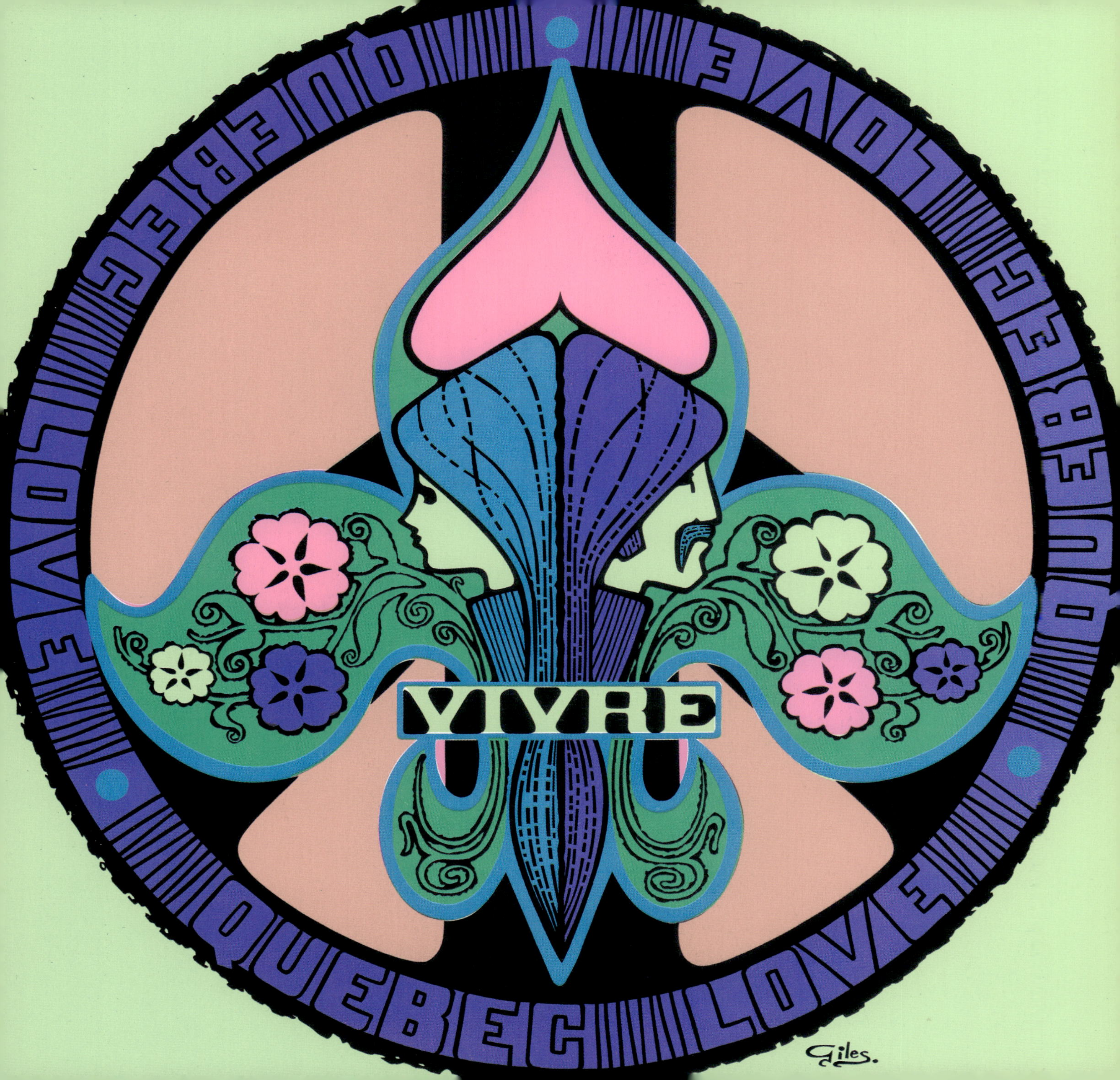
QUEBEC
LOVE
VIVRE
Giles.

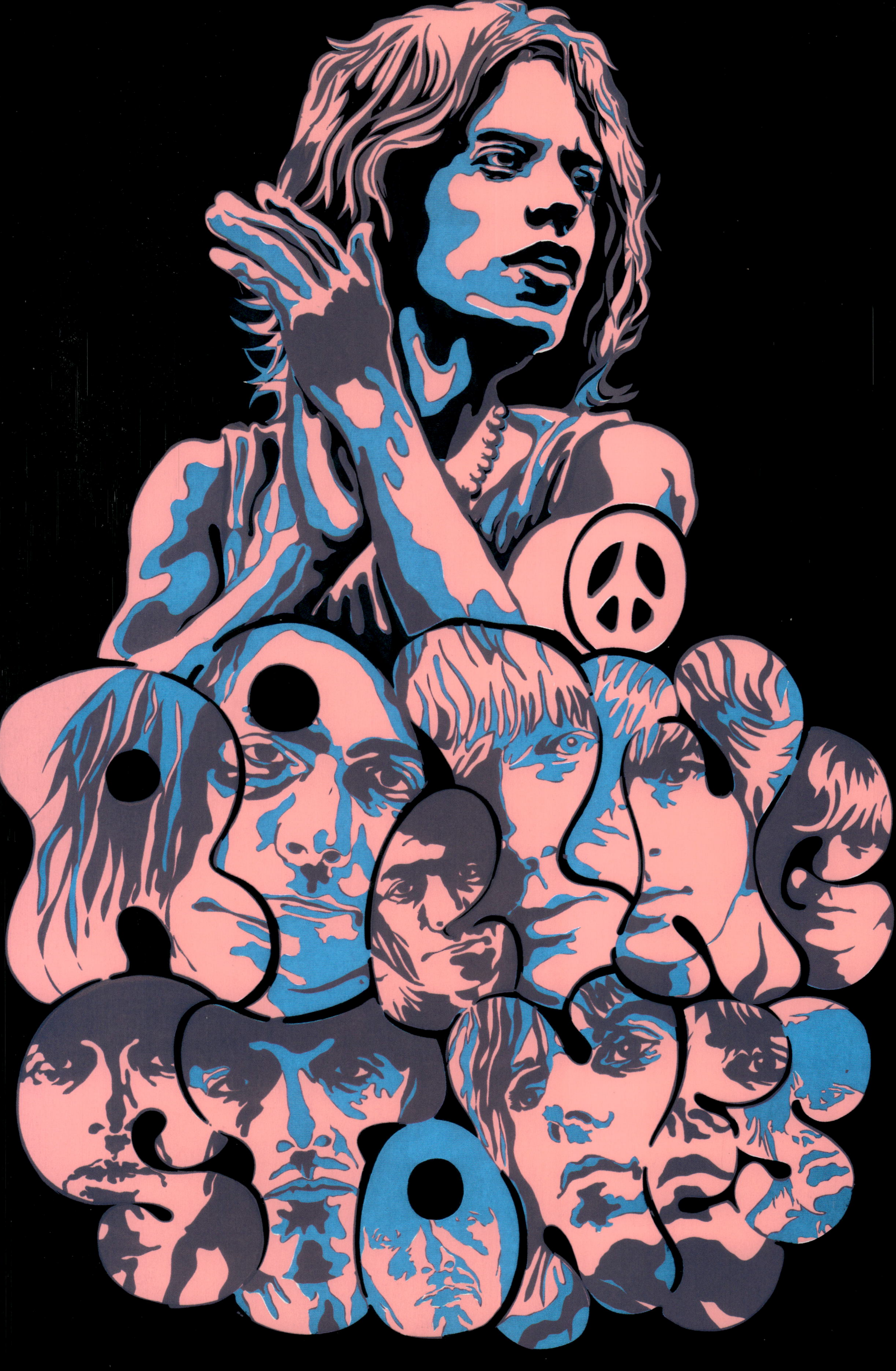

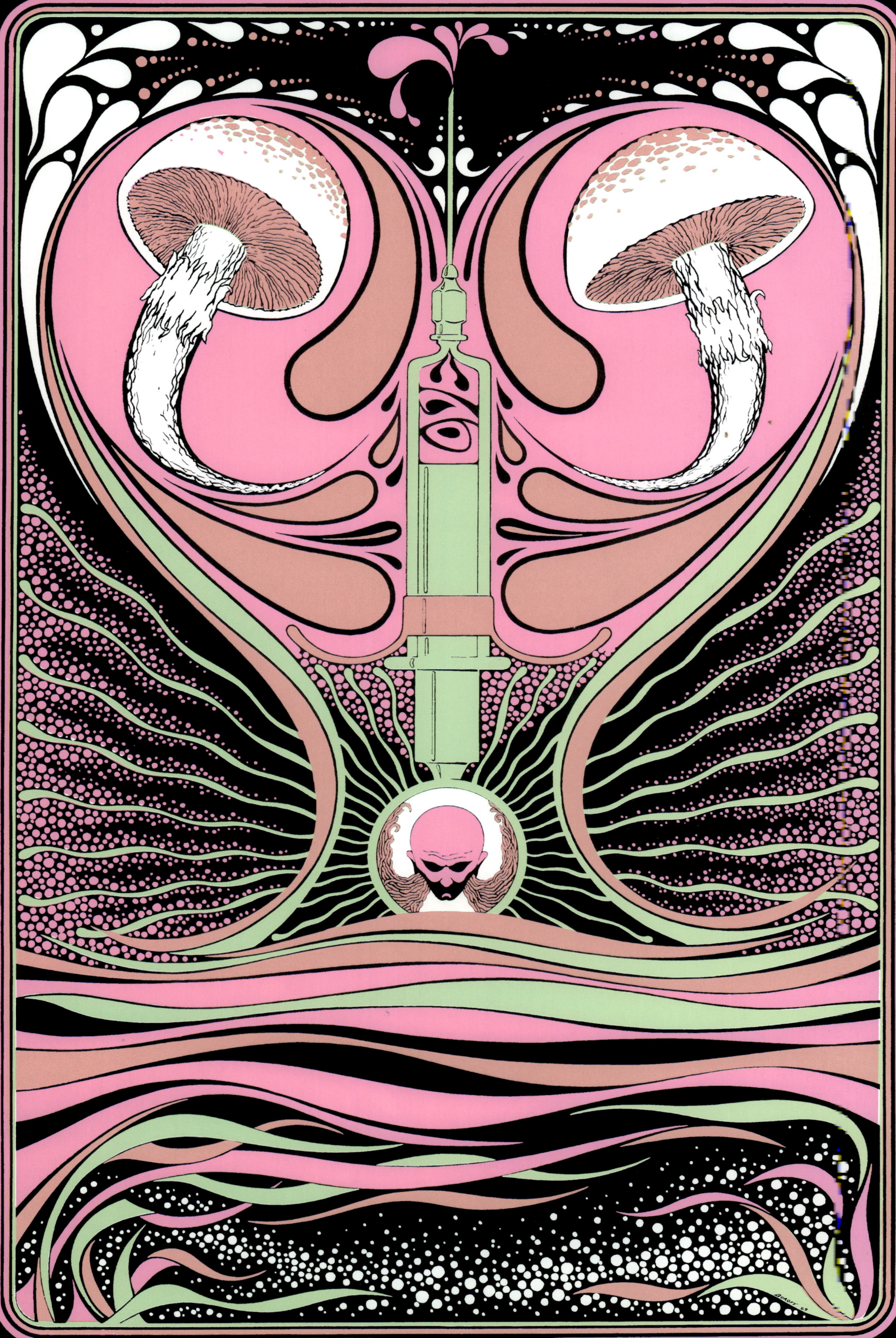

MAINLINE LOVE

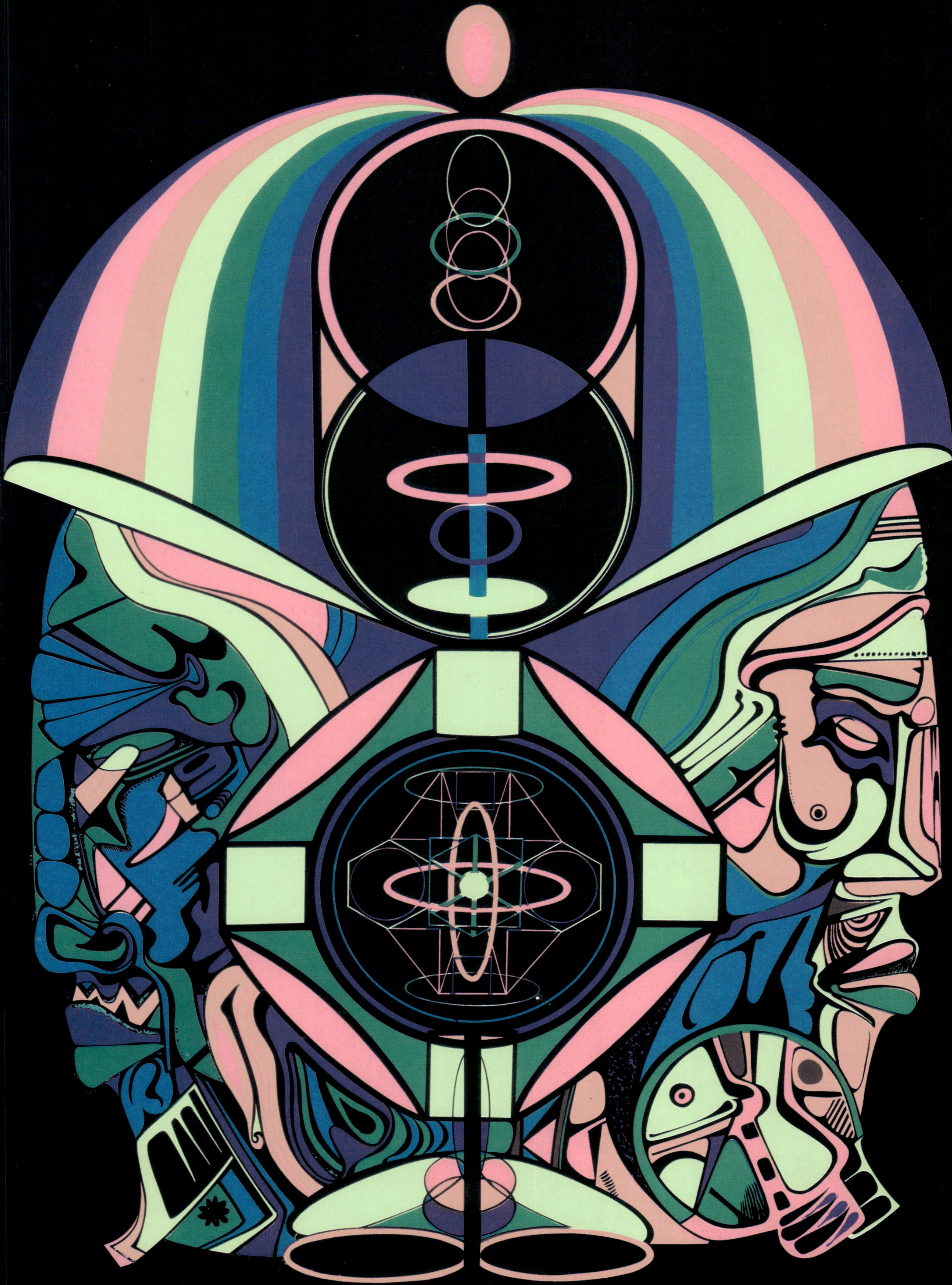

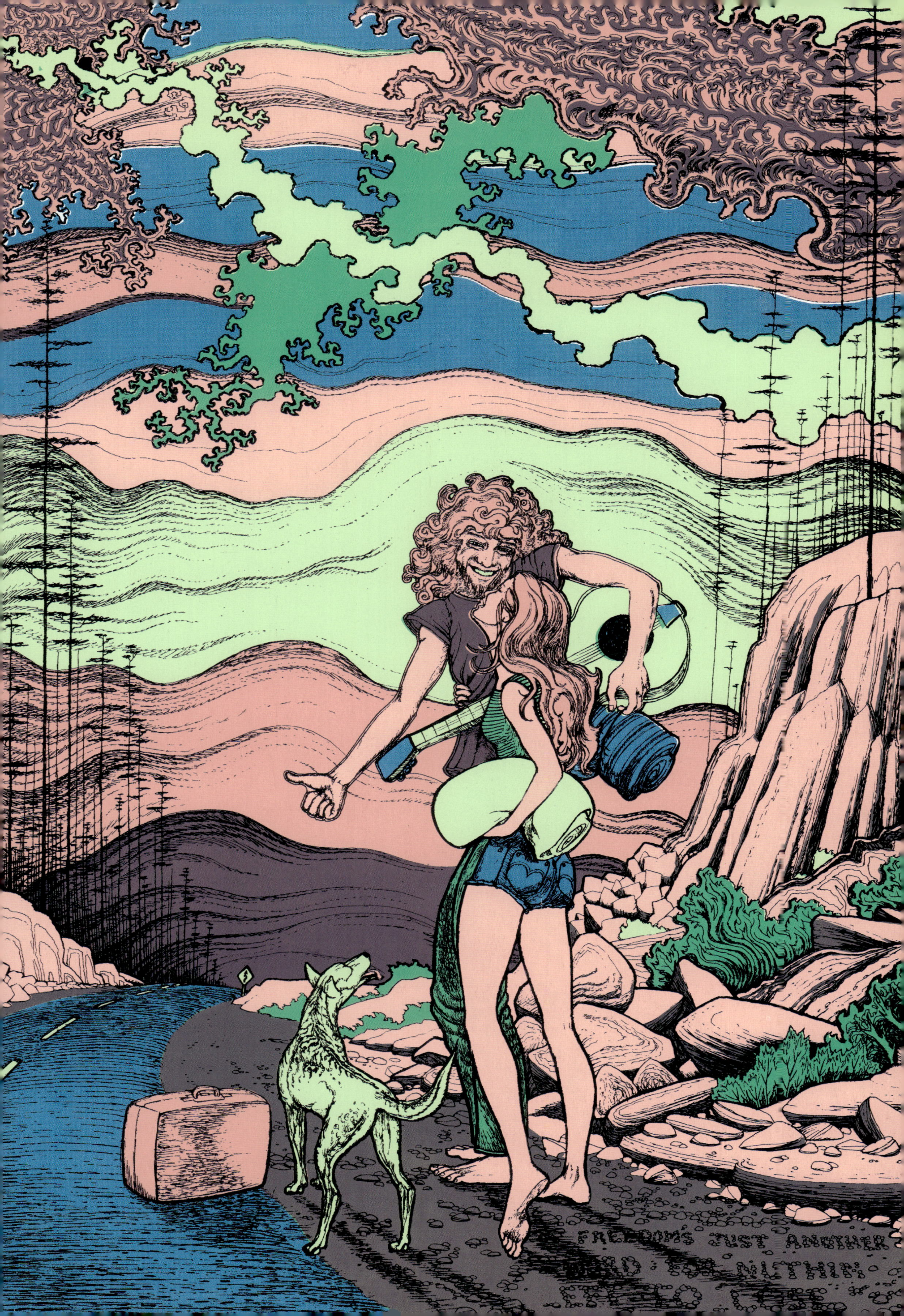
FREEDOM'S JUST ANOTHER
FOR NUTHIN'

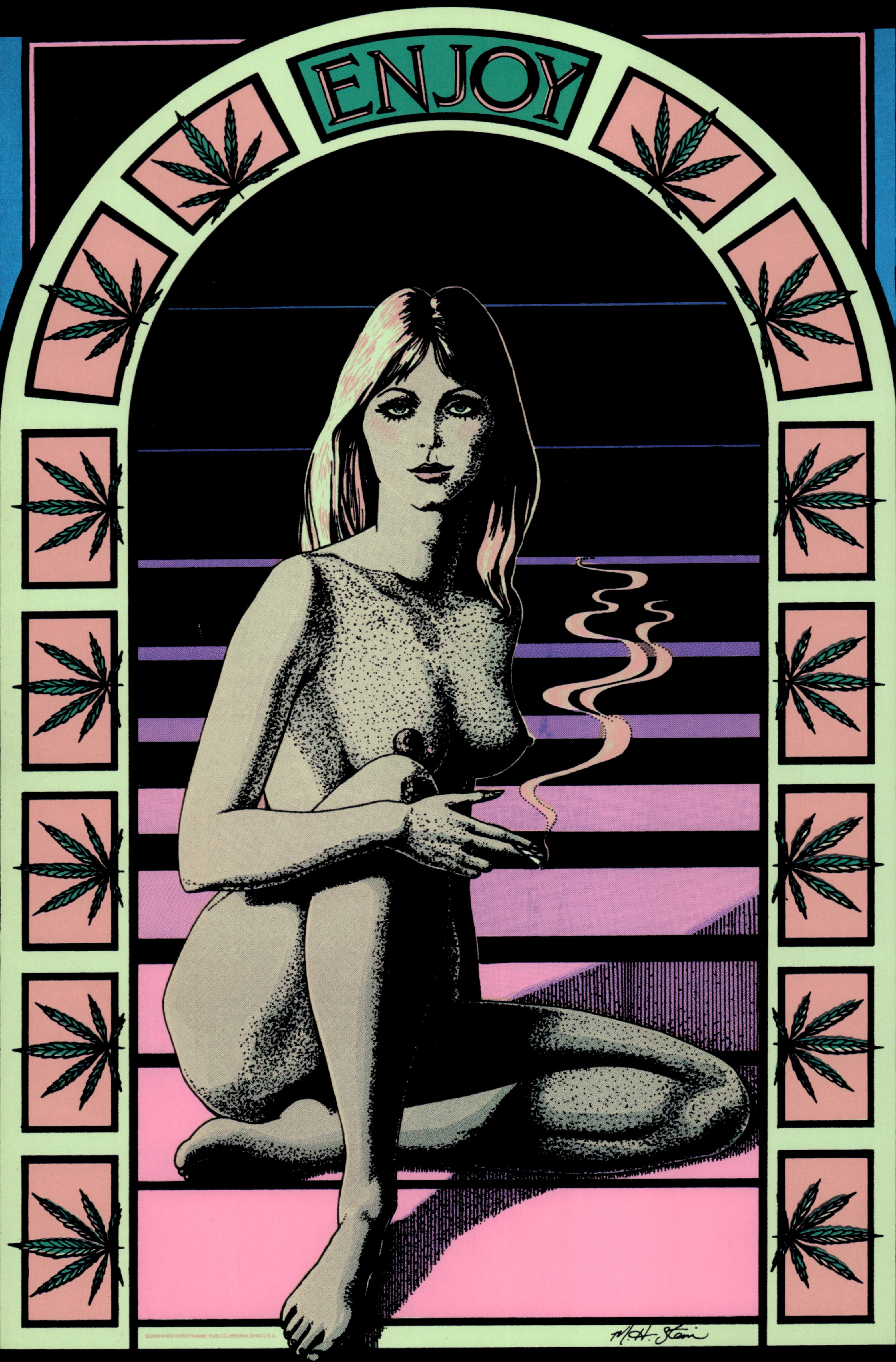
ENJOY
SUNSHINE ©1975 DYNAMIC PUB.CO. MEDINA, OHIO U.S.A.

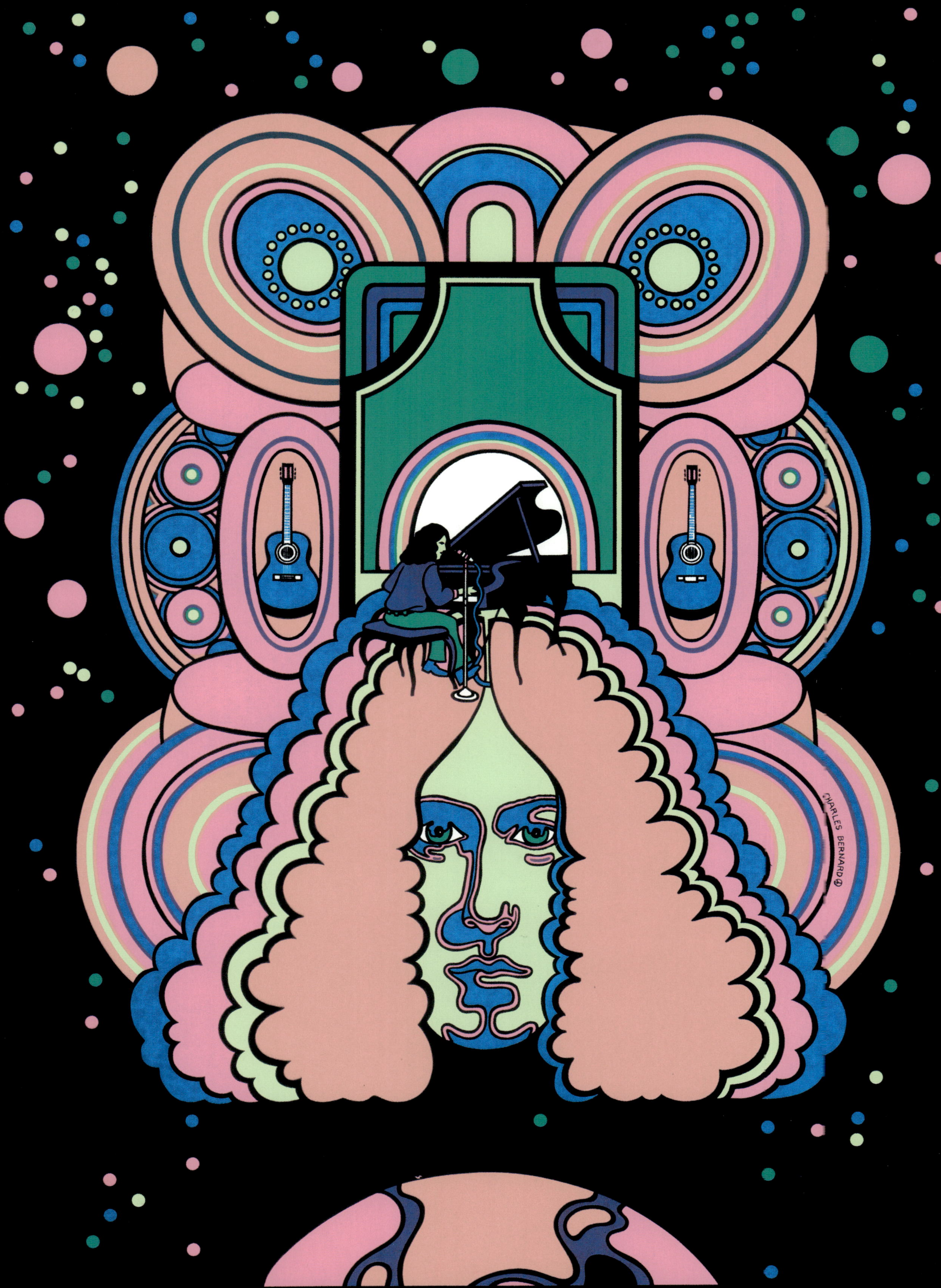
CHARLES BERNARD ©

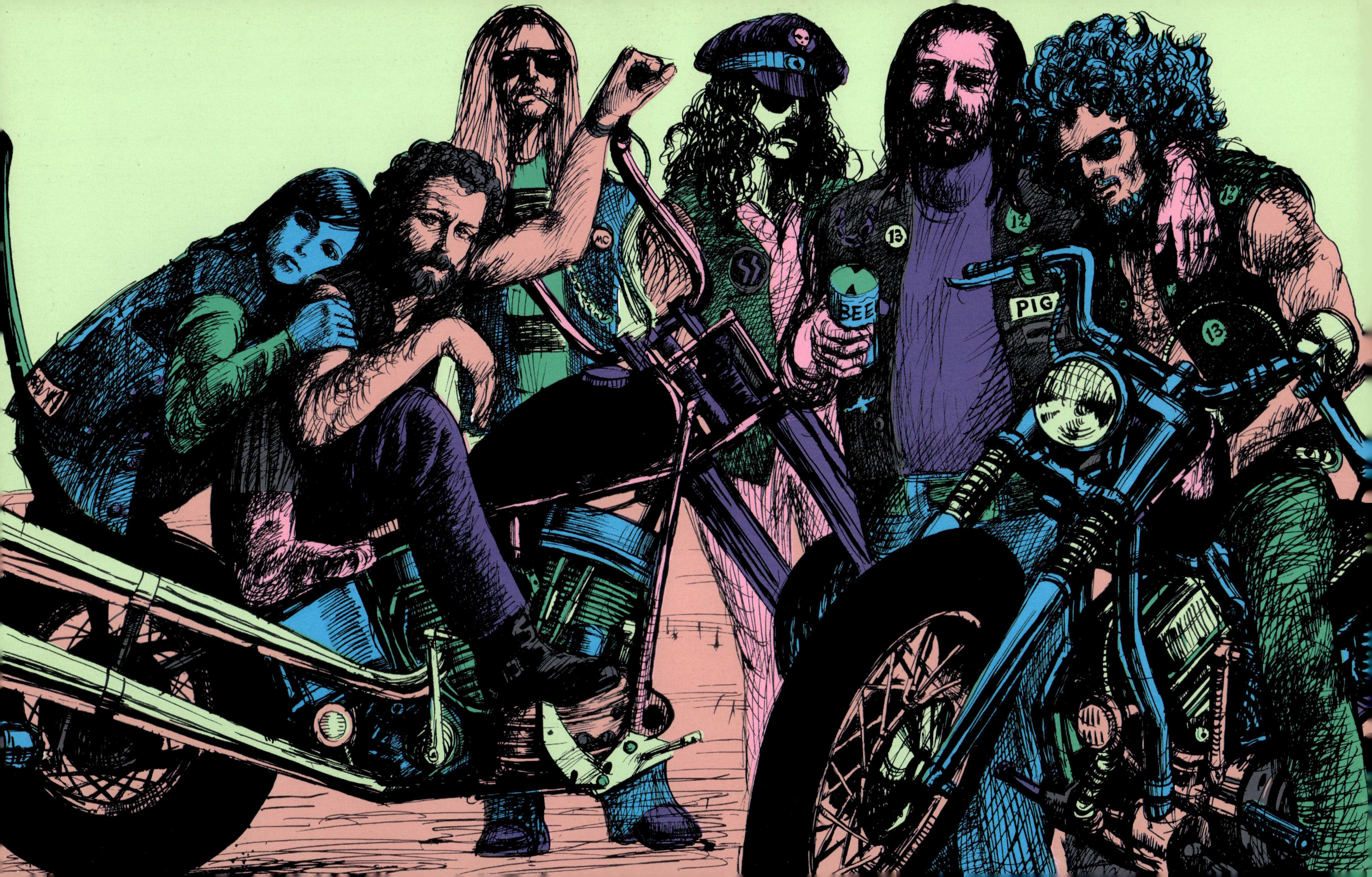
PIG
BEE

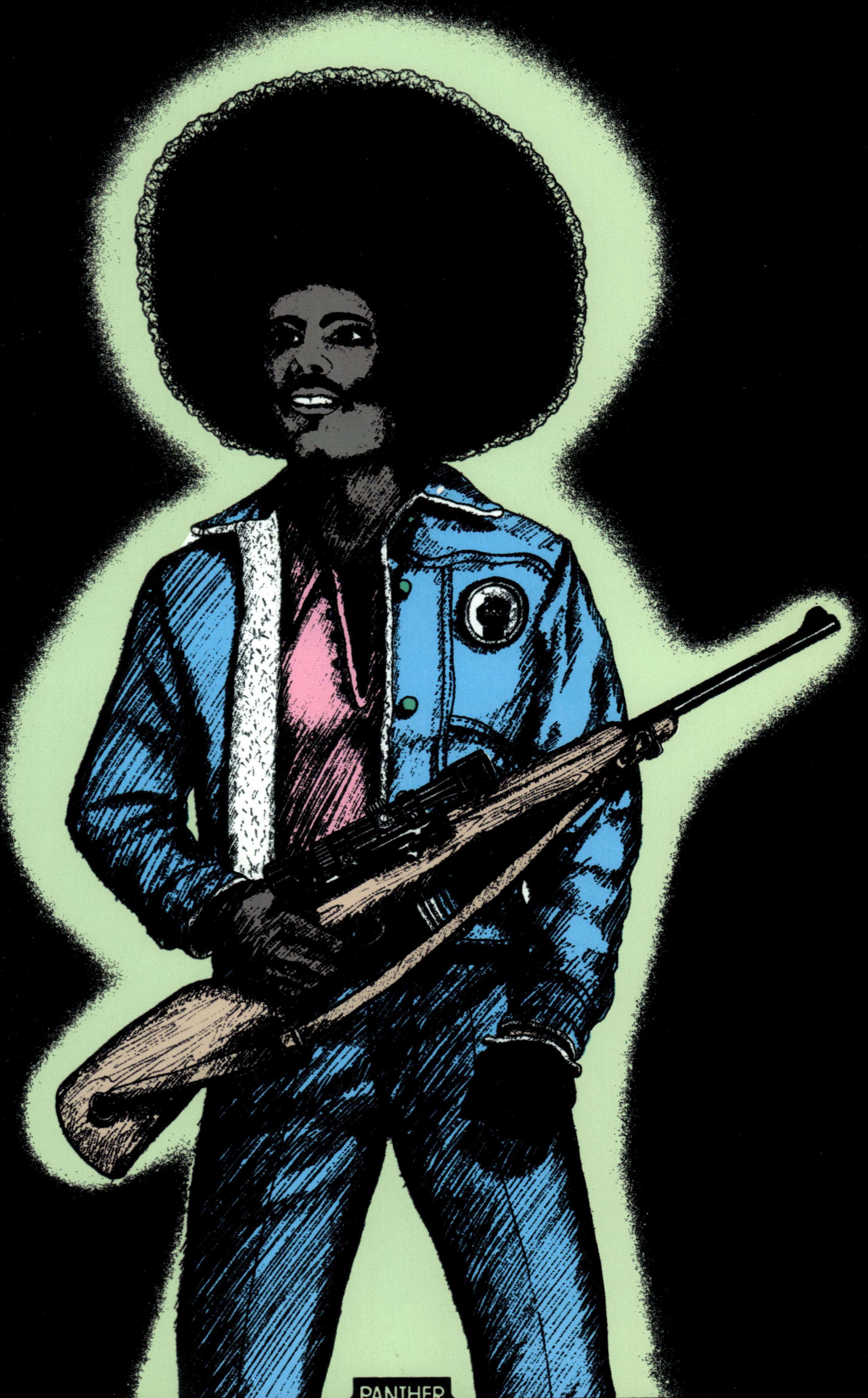
PANTHER

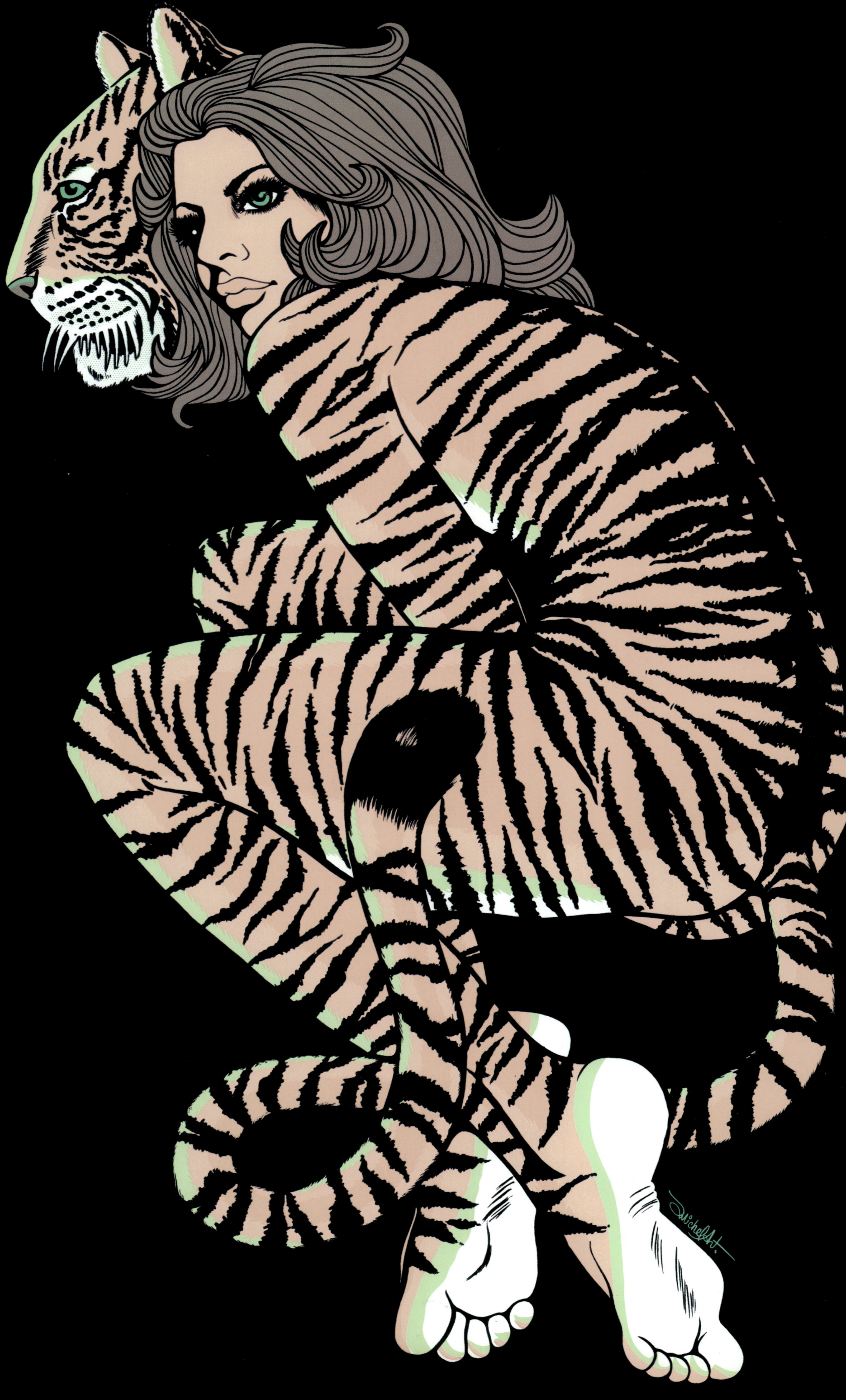

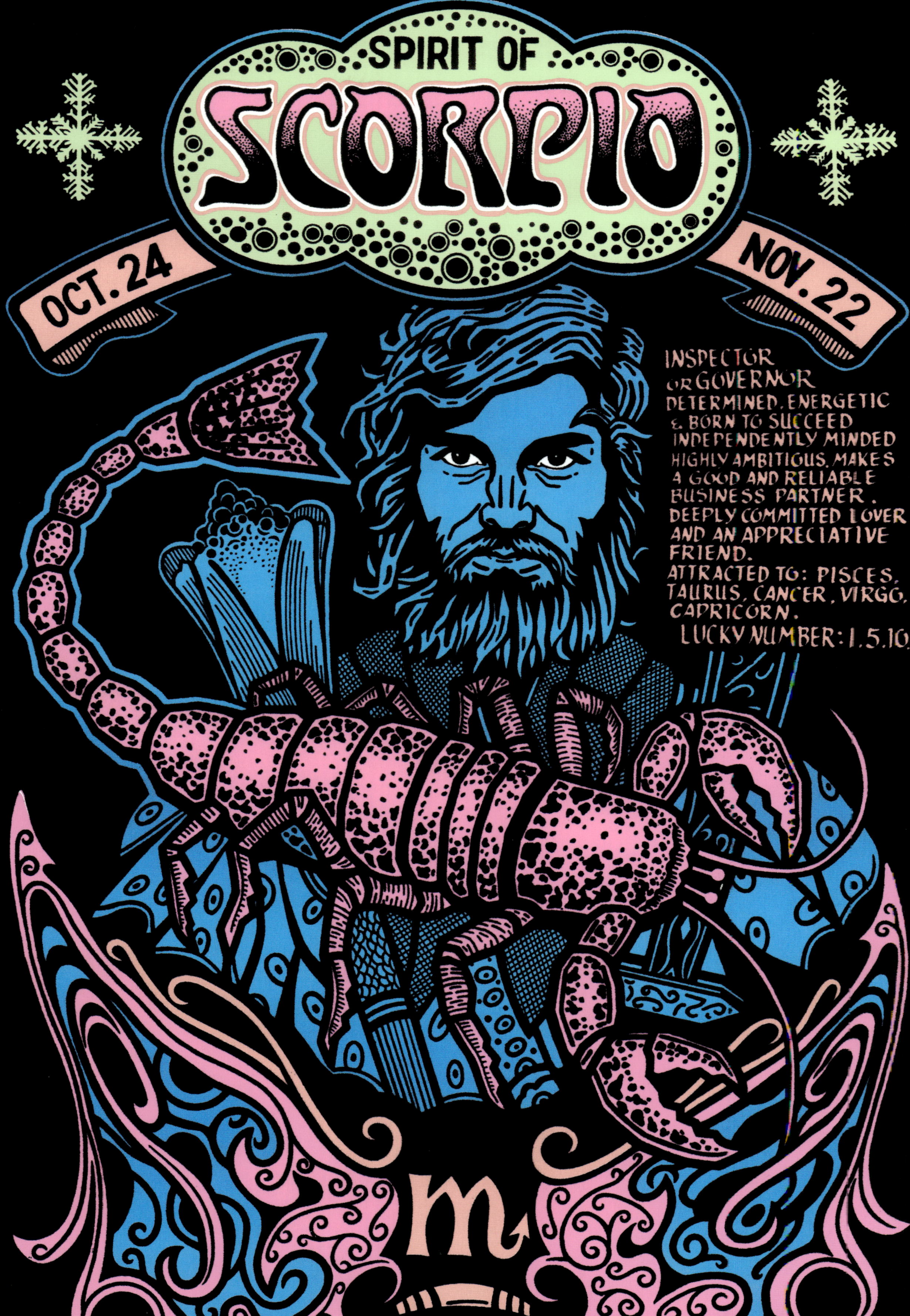
SPIRIT OF
SCORPIO
OCT. 24
NOV. 22
INSPECTOR
OR GOVERNOR
DETERMINED, ENERGETIC
& BORN TO SUCCEED
INDEPENDENTLY MINDED
HIGHLY AMBITIOUS, MAKES
A GOOD AND RELIABLE
BUSINESS PARTNER.
DEEPLY COMMITTED LOVER
AND AN APPRECIATIVE
FRIEND.
ATTRACTED TO: PISCES,
TAURUS, CANCER, VIRGO,
CAPRICORN.
LUCKY NUMBER: 1. 5. 10.

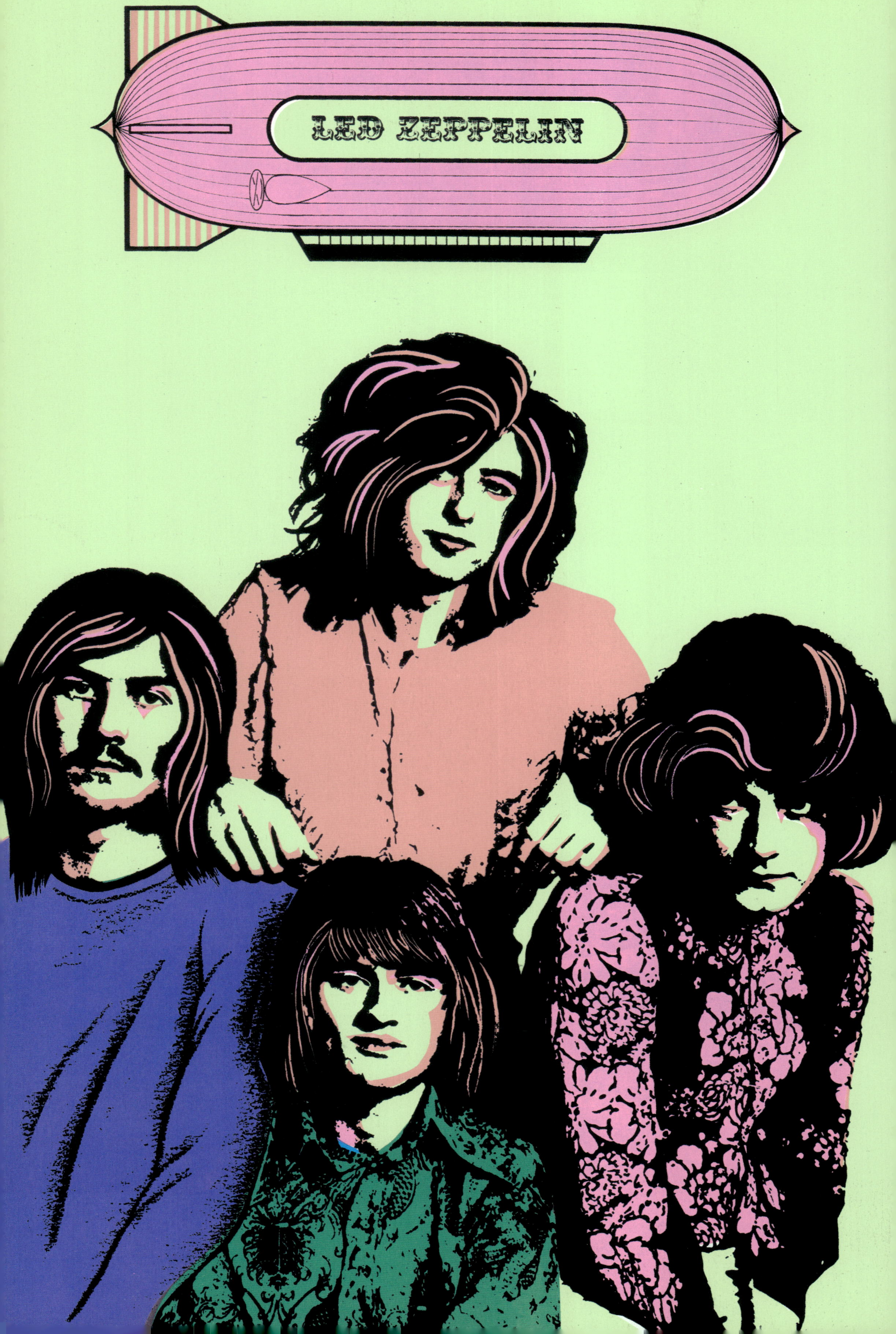
LED ZEPPELIN

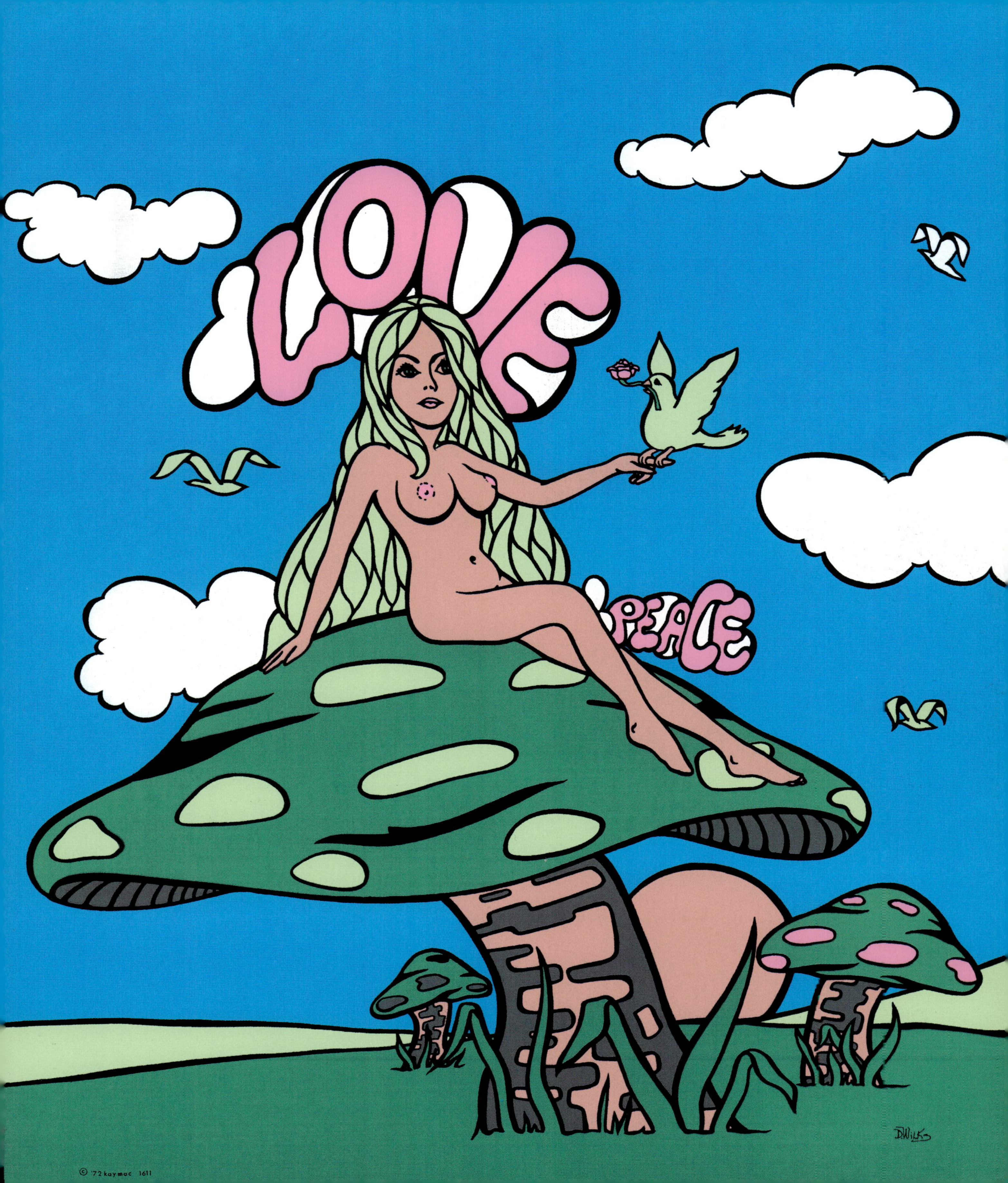
LOVE
PEACE
D.Wilks
© '72 kaymac 1611

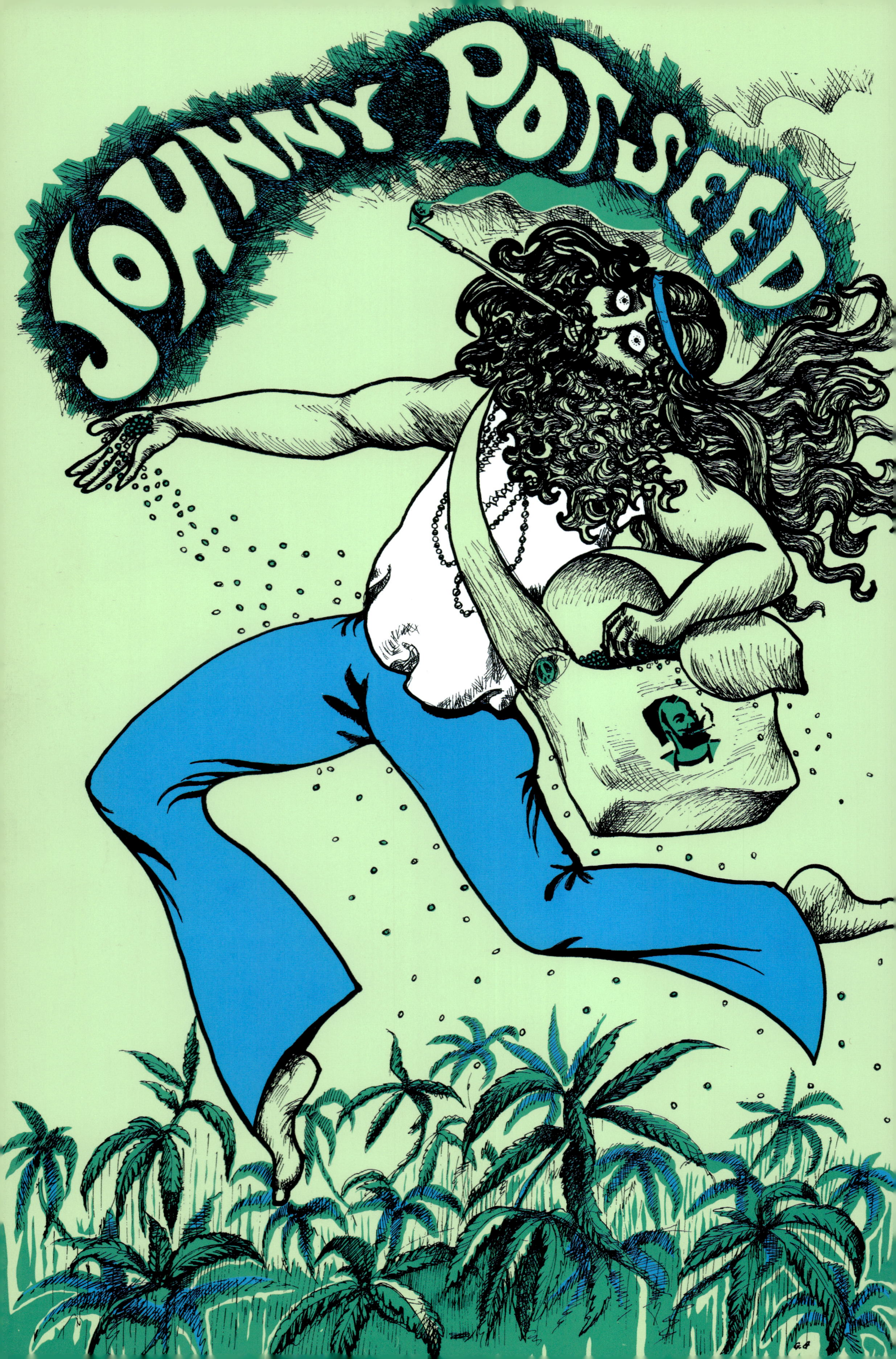
JOHNNY POTSEED

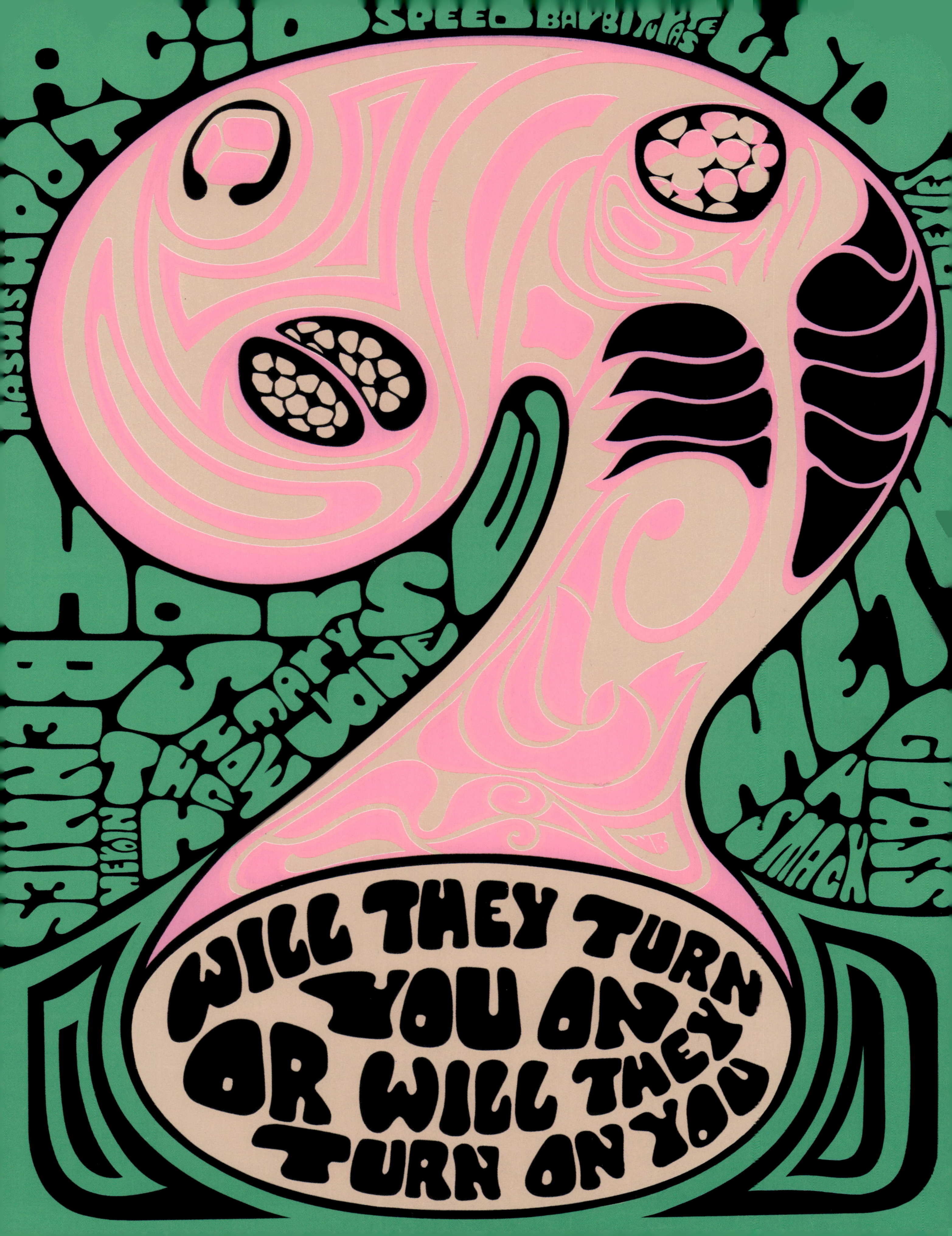
ACID
SPEED
LSD
MARY JANE
SMACK
WILL THEY TURN YOU ON OR WILL THEY TURN ON YOU

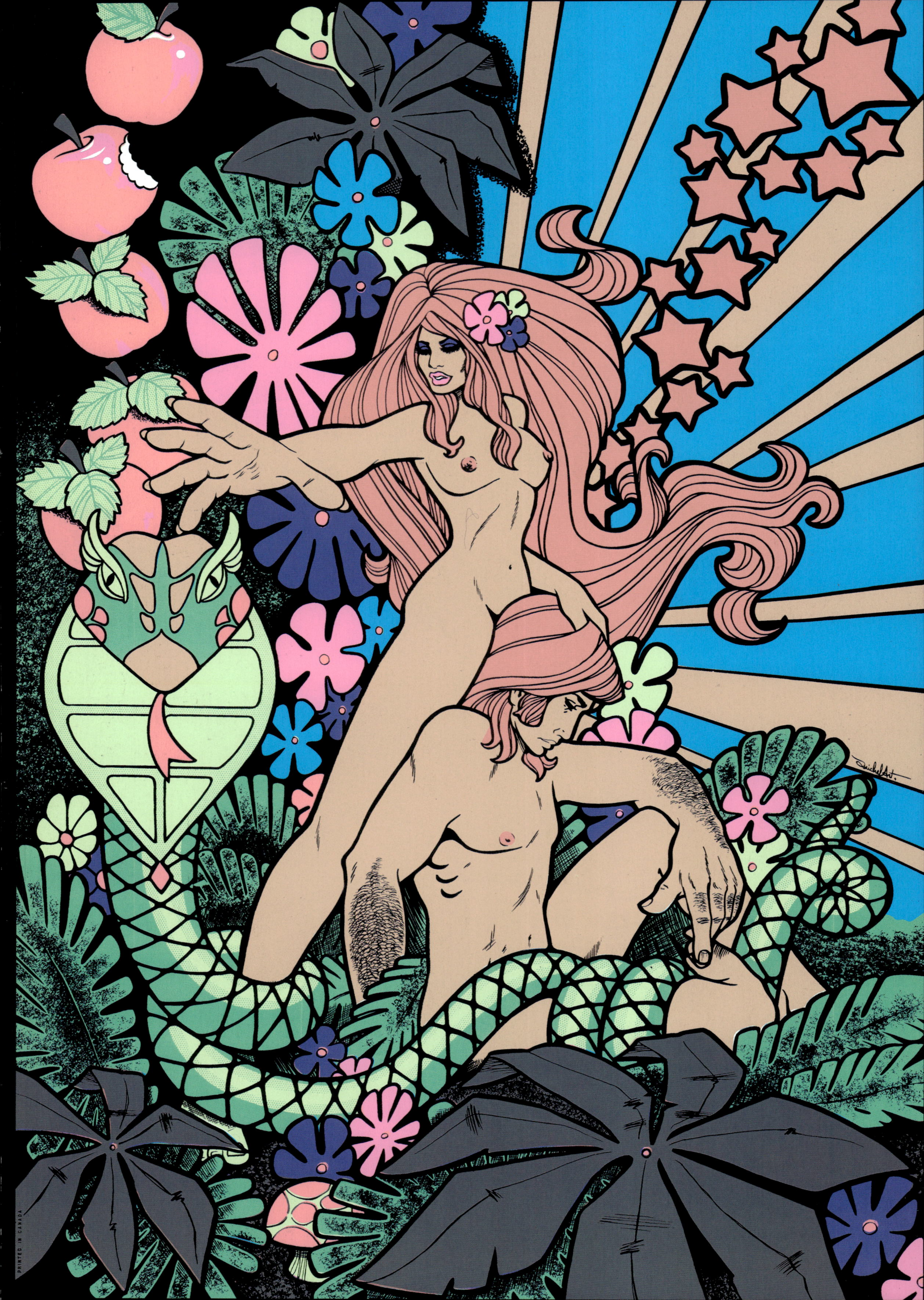
PRINTED IN CANADA

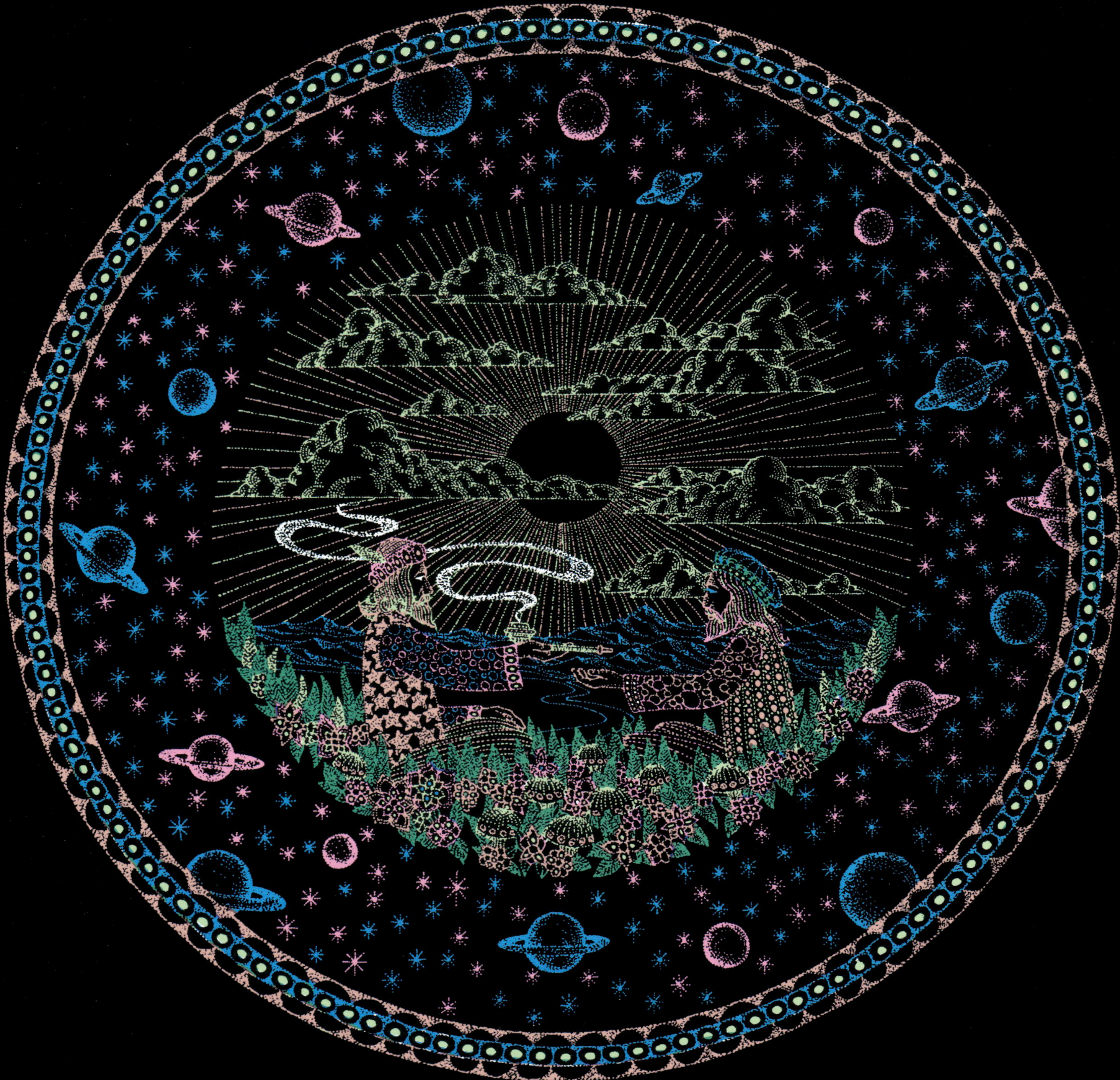

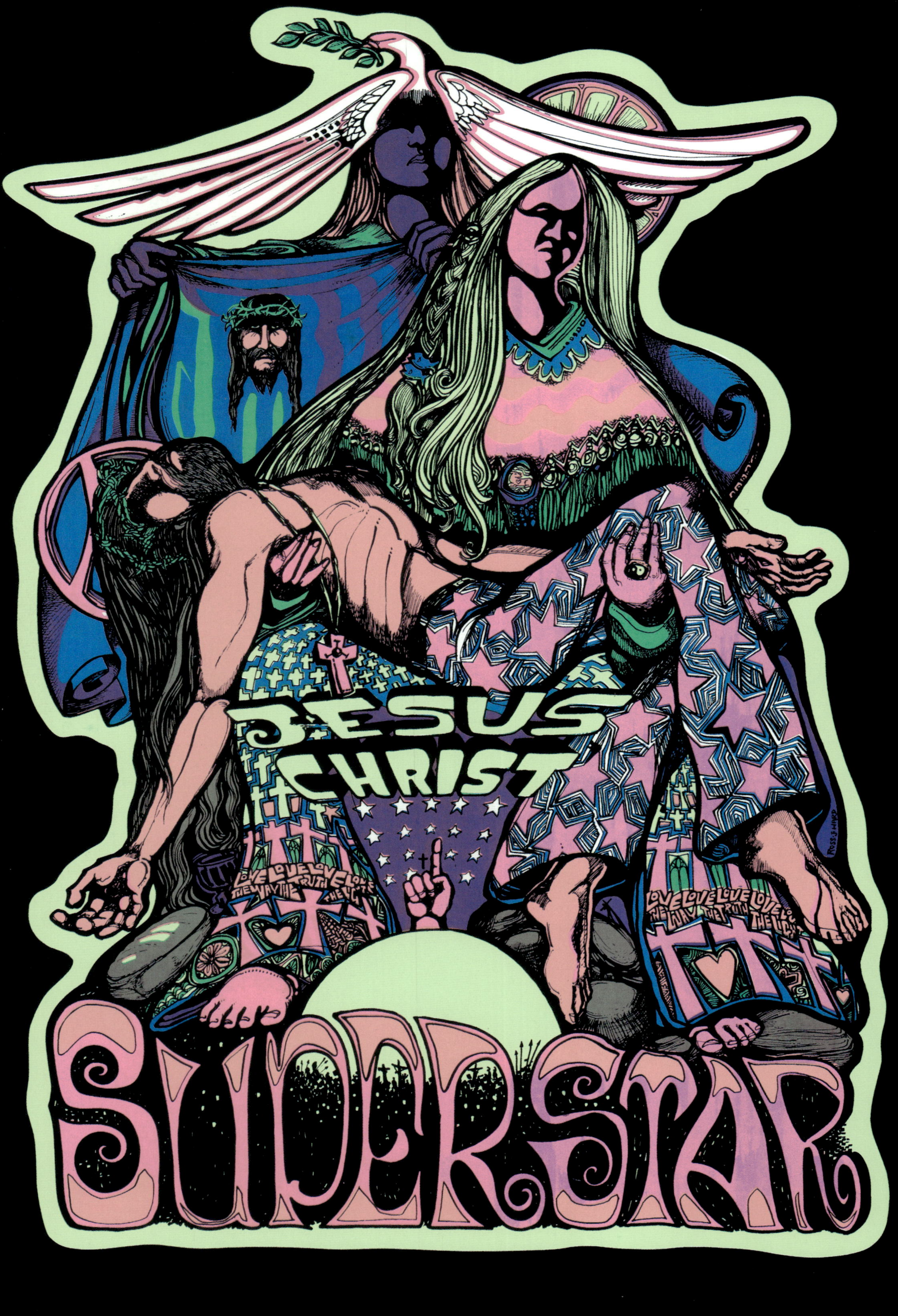
JESUS
CHRIST
SUPERSTAR

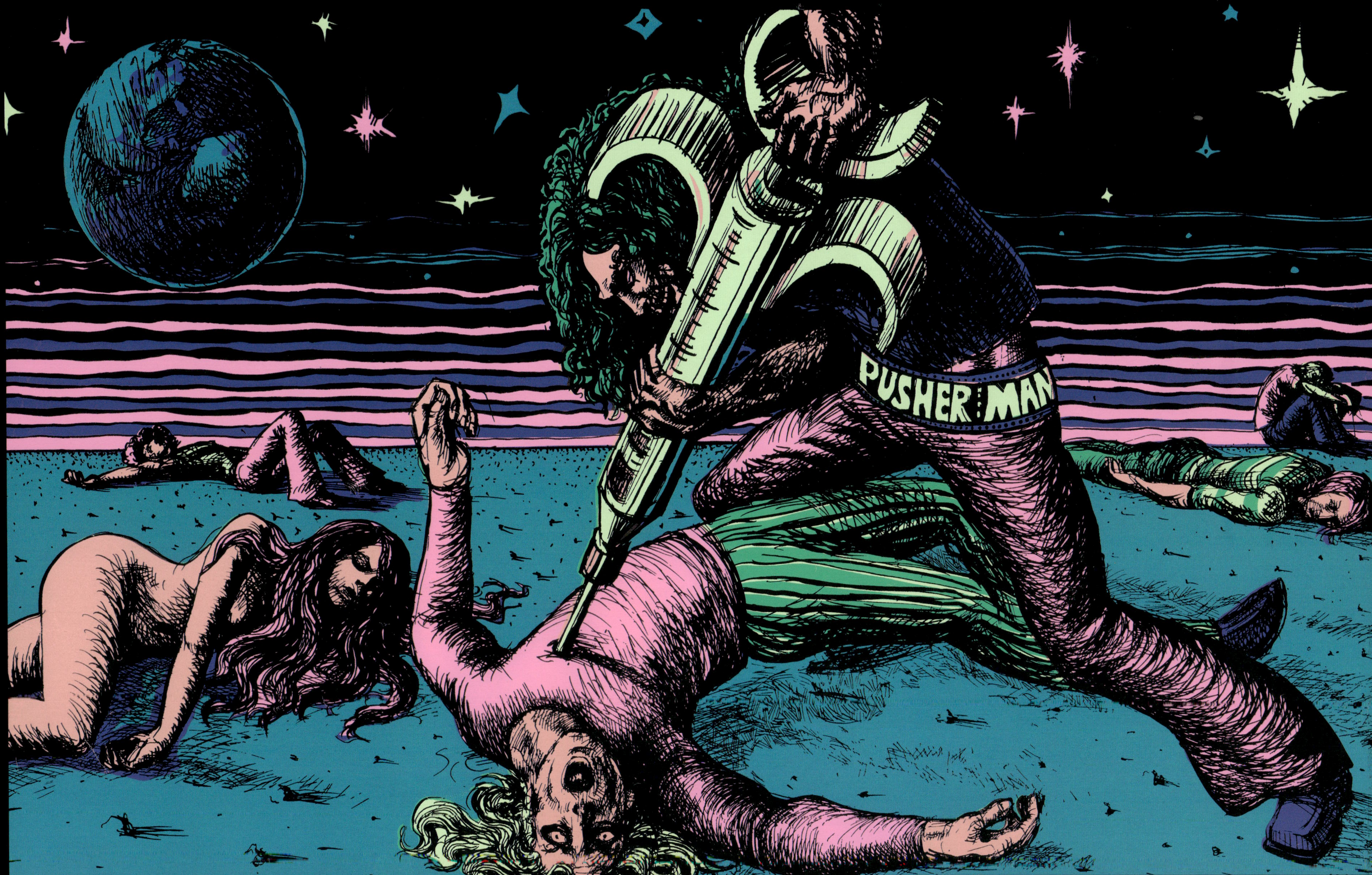
PUSHER MAN

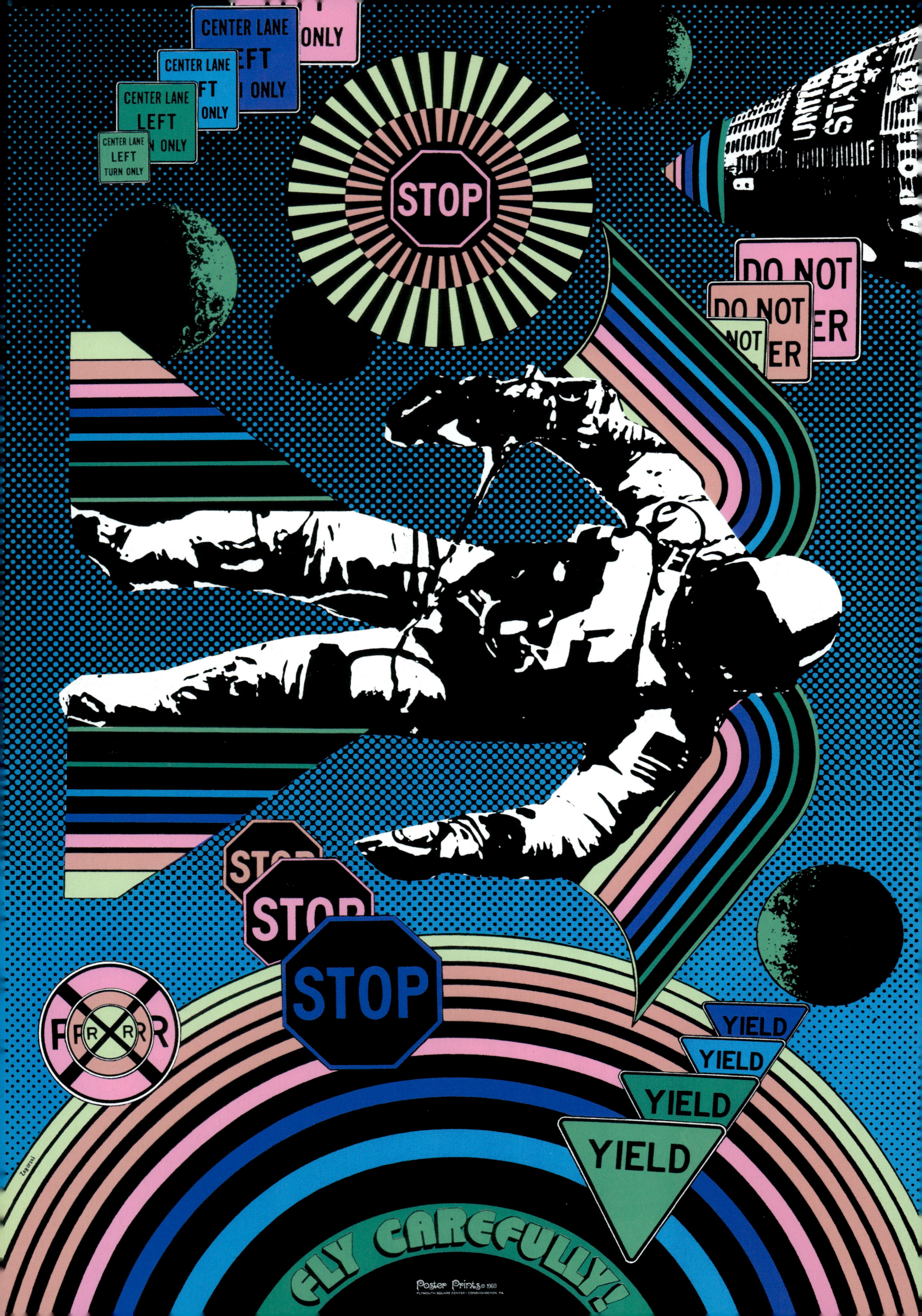
CENTER LANE
LEFT
TURN ONLY
STOP
DO NOT
STOP
STOP
STOP
YIELD
YIELD
YIELD
YIELD
FLY CAREFULLY!
Zagorski
Poster Prints
PLYMOUTH SQUARE CENTER • CONSHOHOCKEN, PA

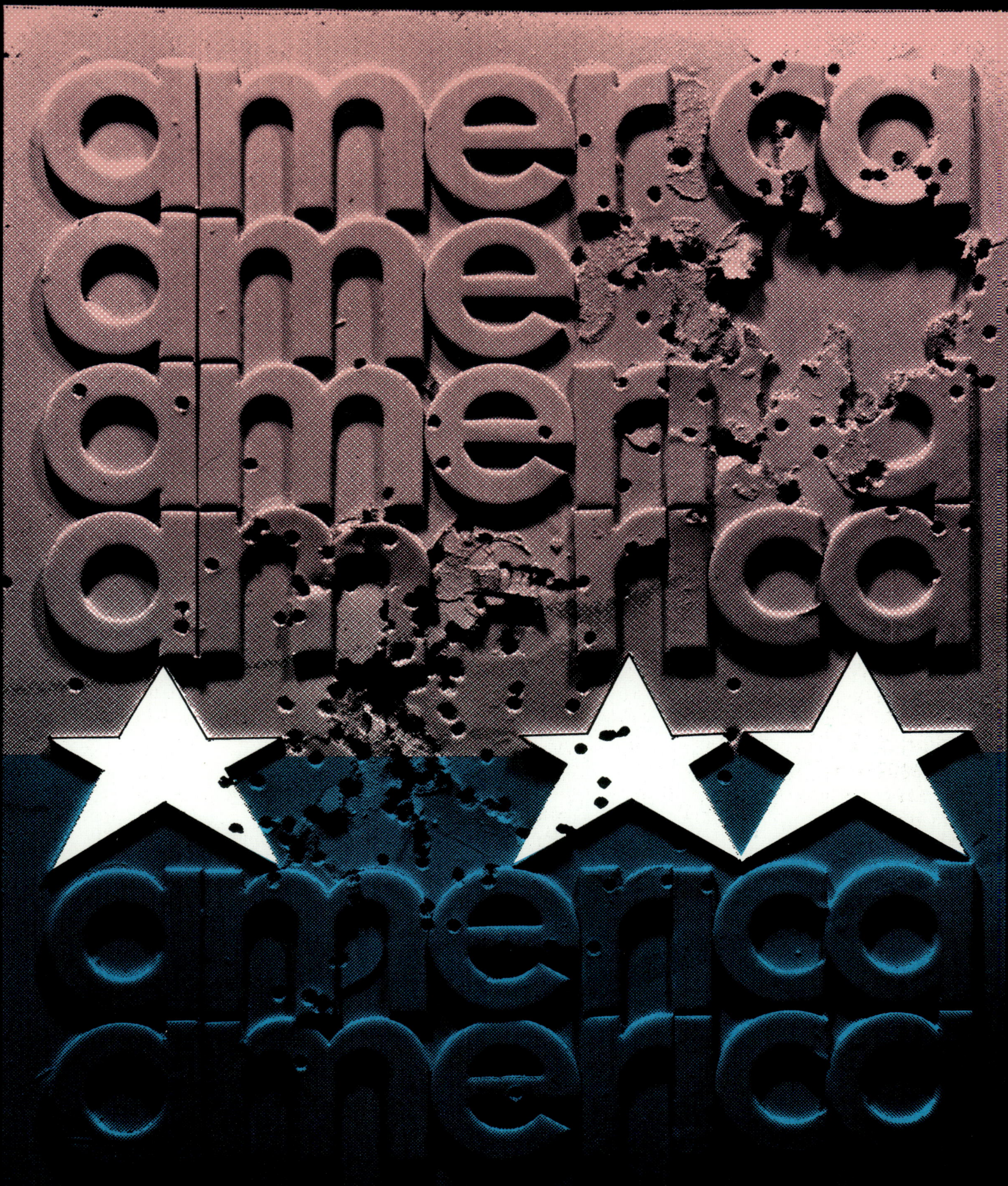
america
america
america

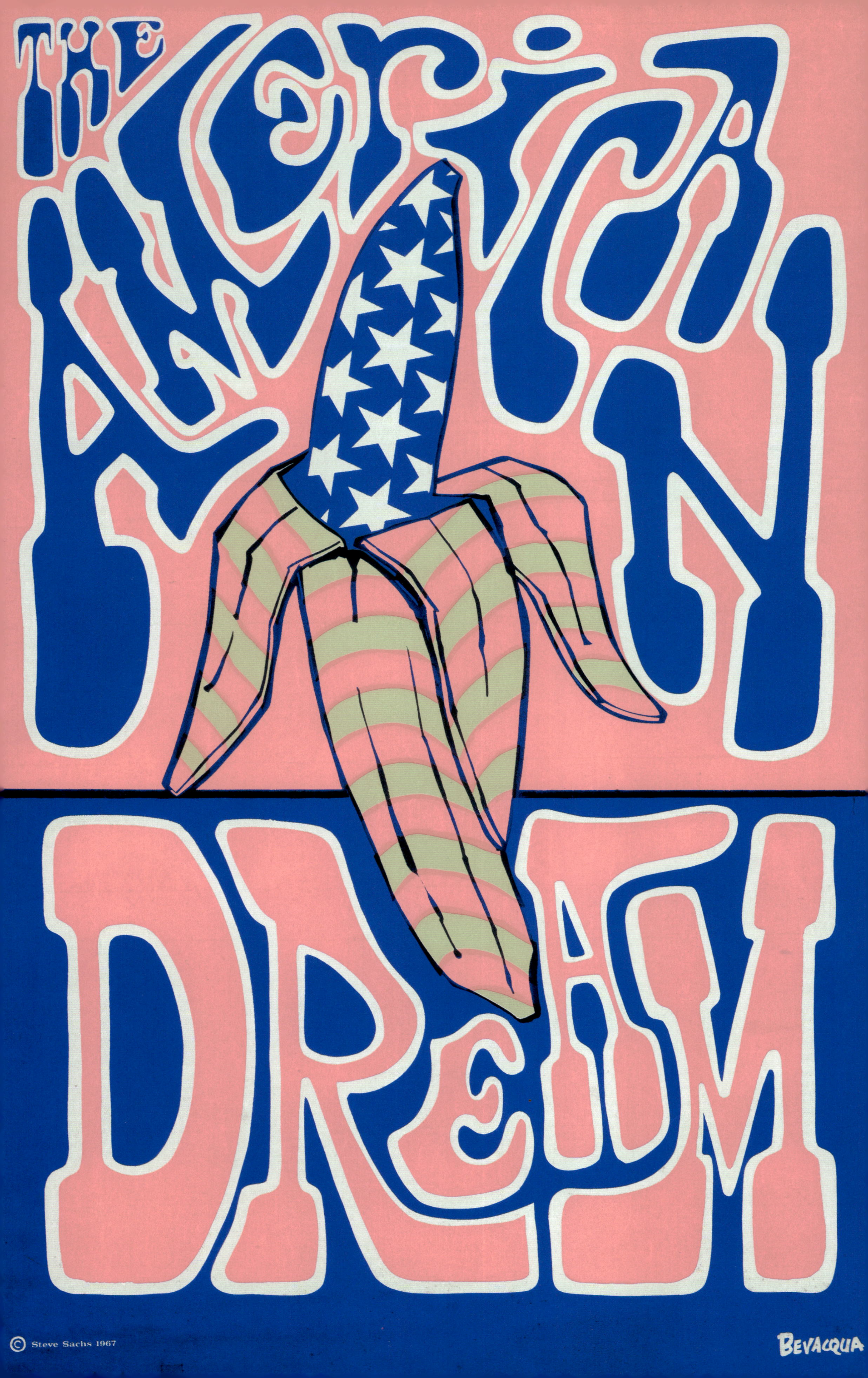
THE AMERICAN
DREAM
© Steve Sachs 1967
BEVACQUA

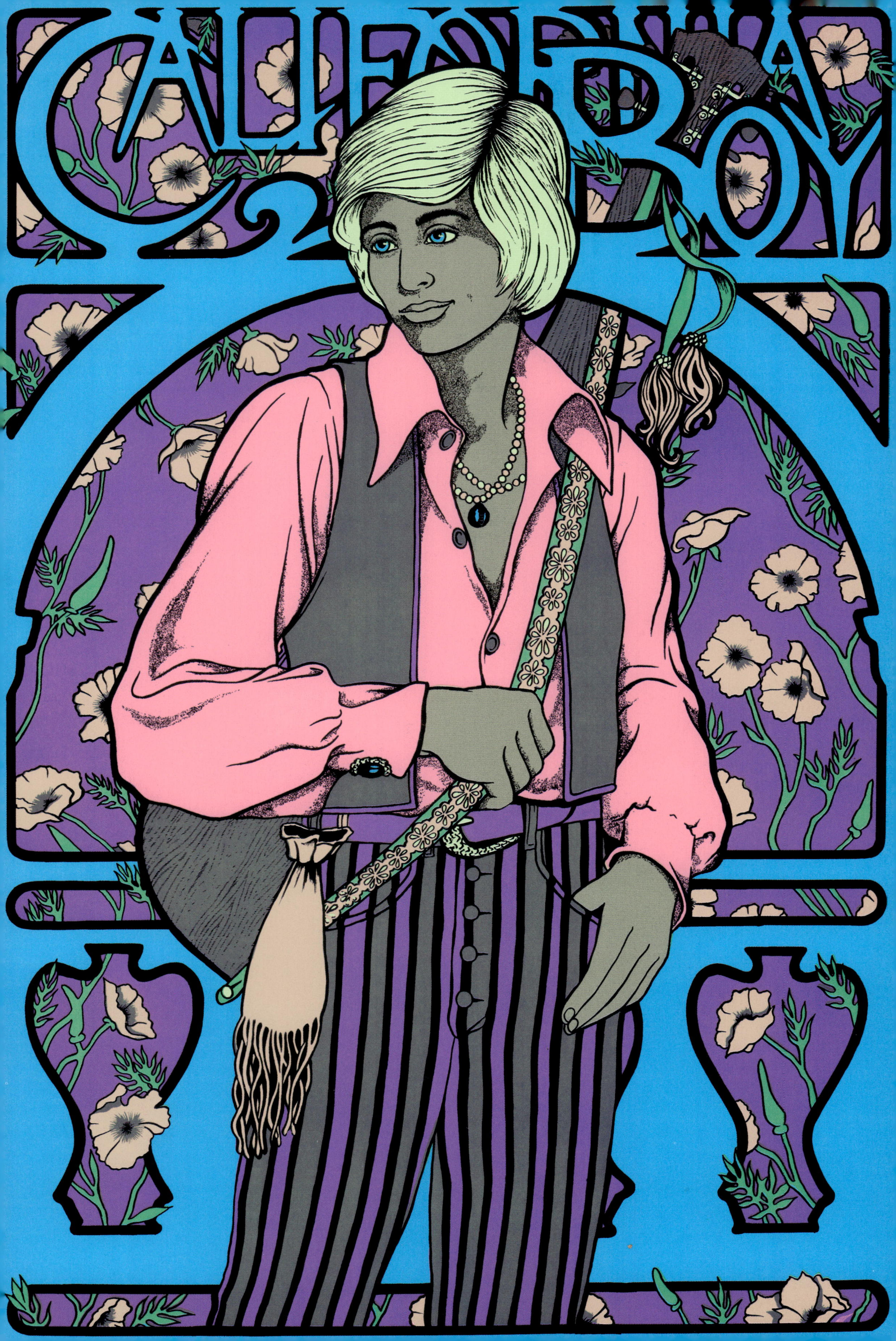
CALIFORNIA BOY

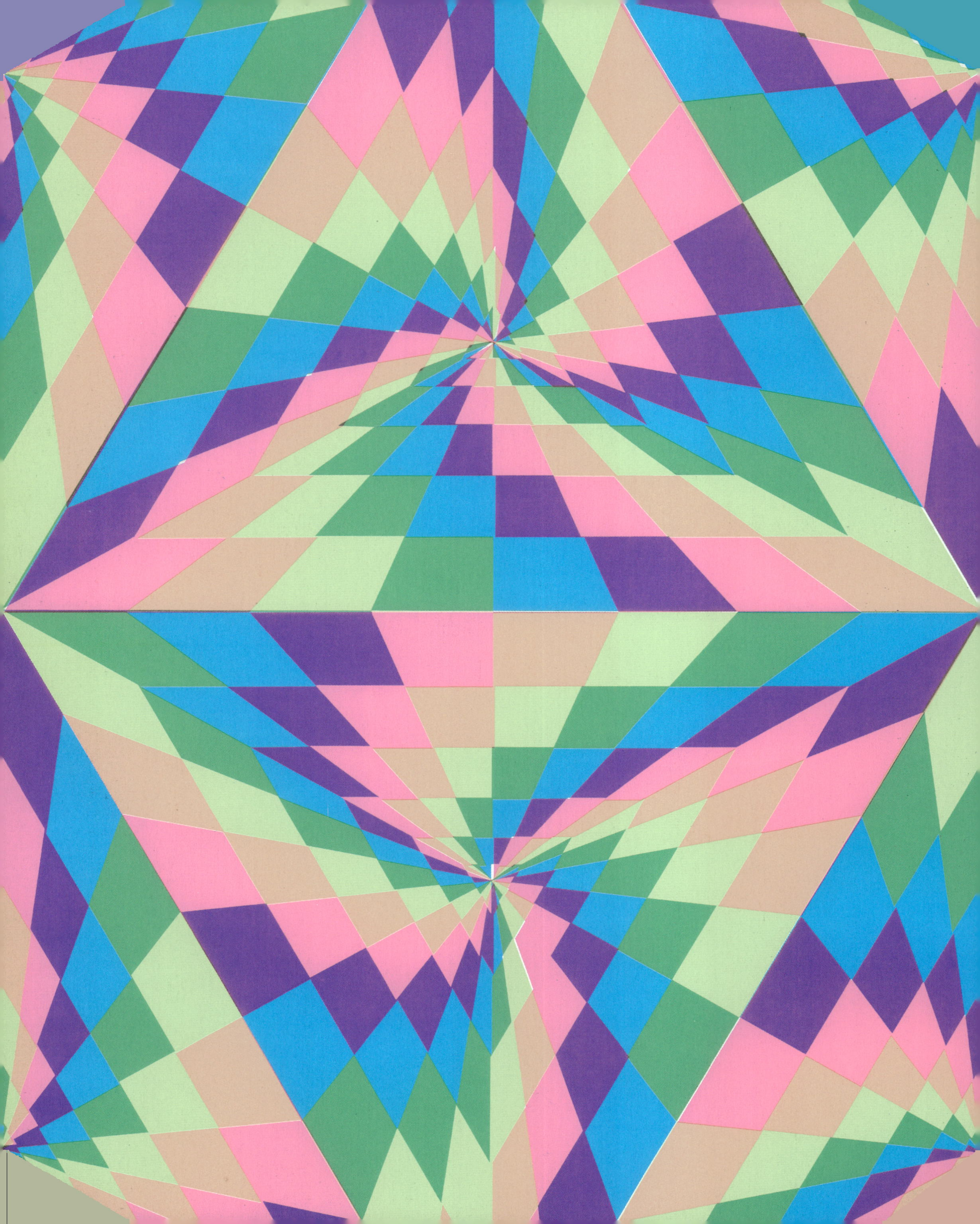

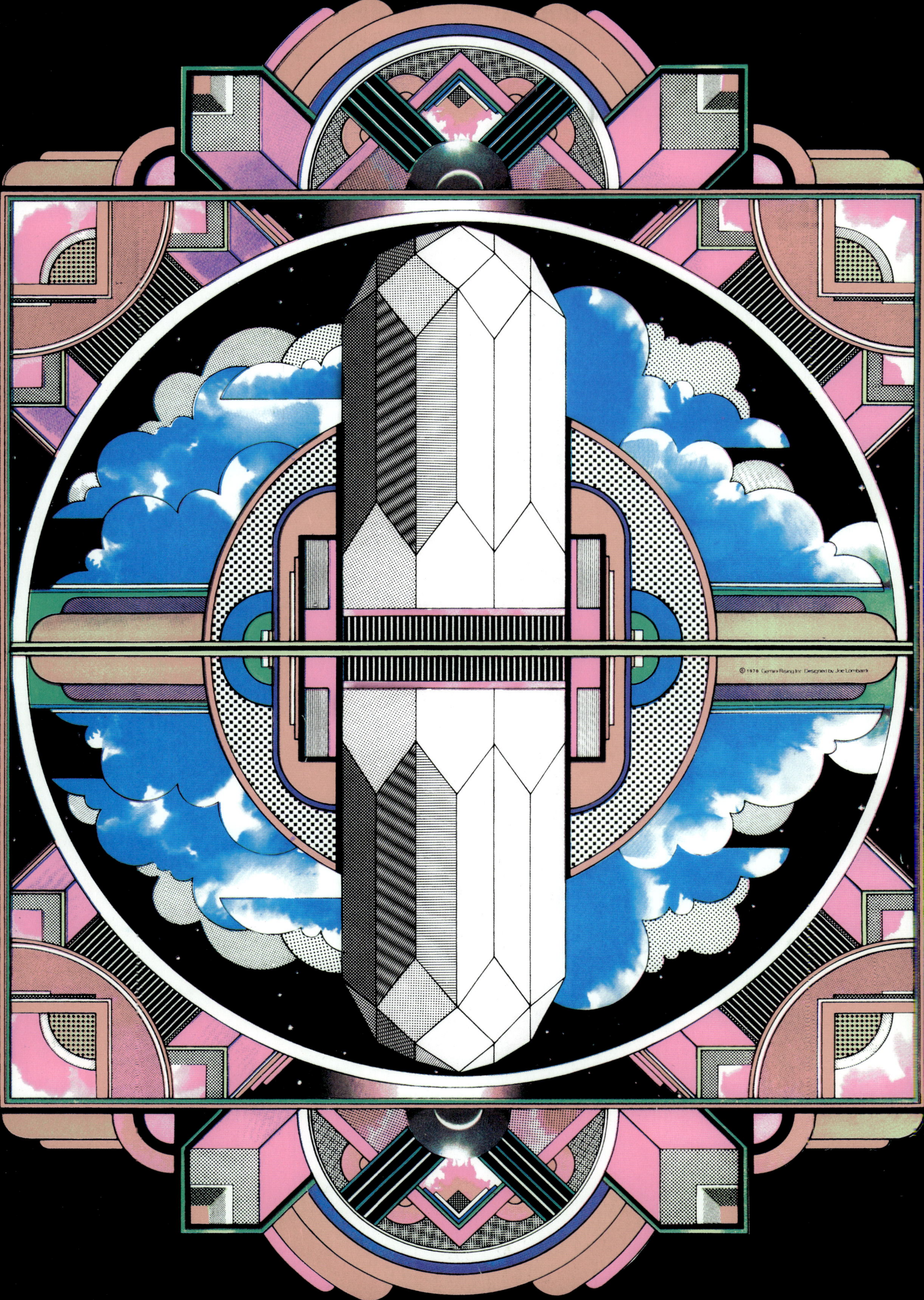
© 1970 Gemini Rising Inc Designed by Joe Lombardi

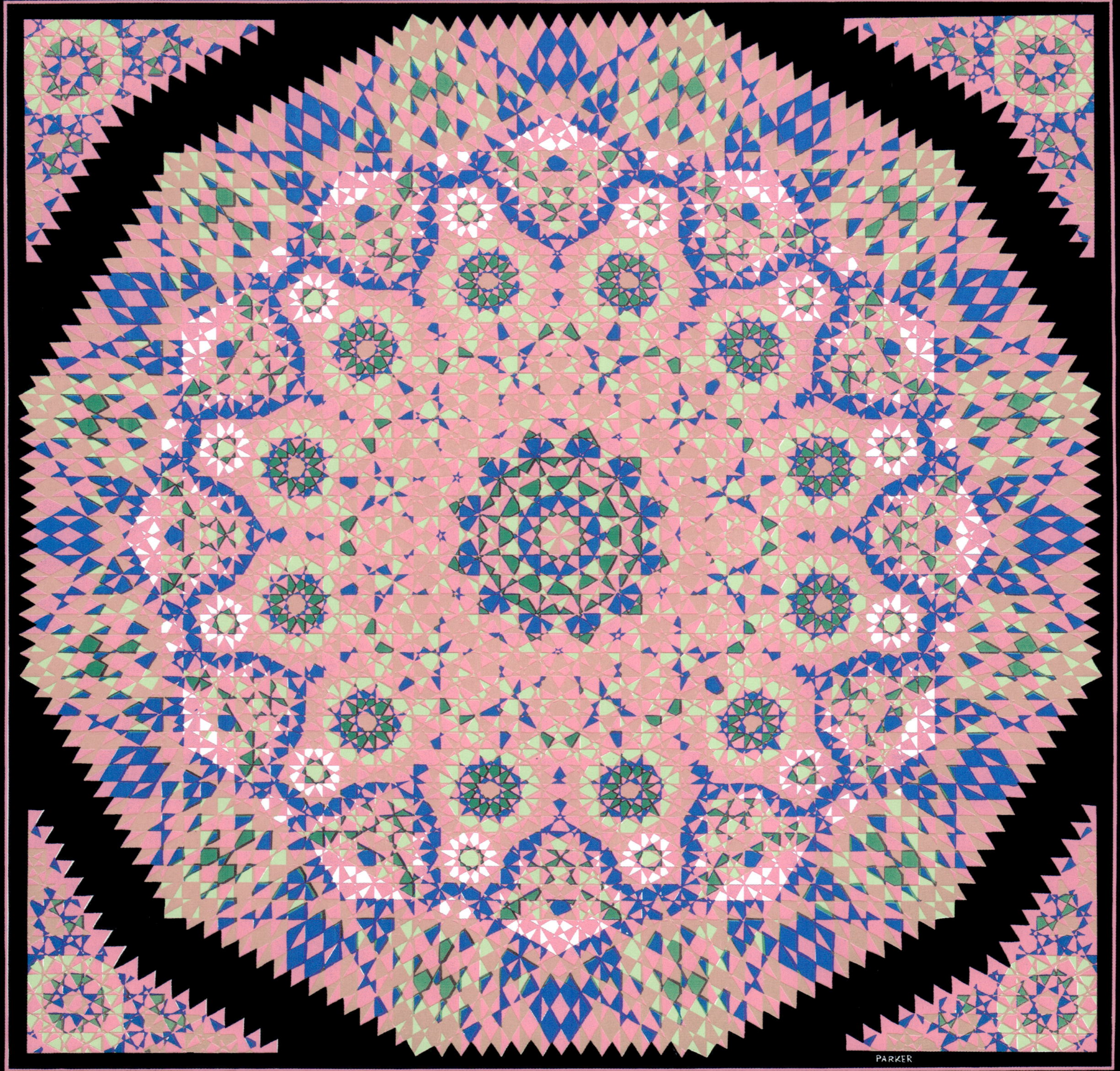
PARKER

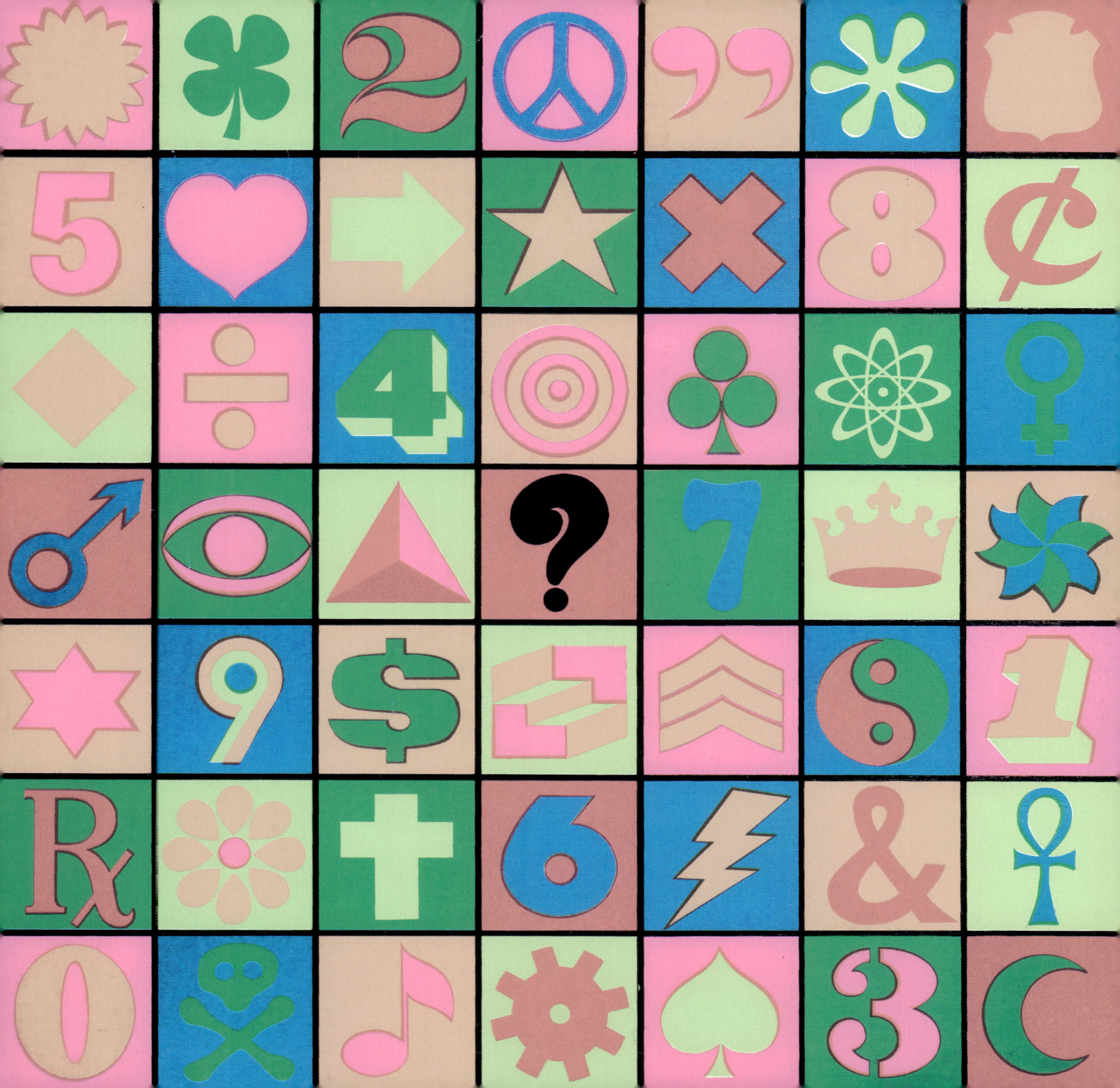

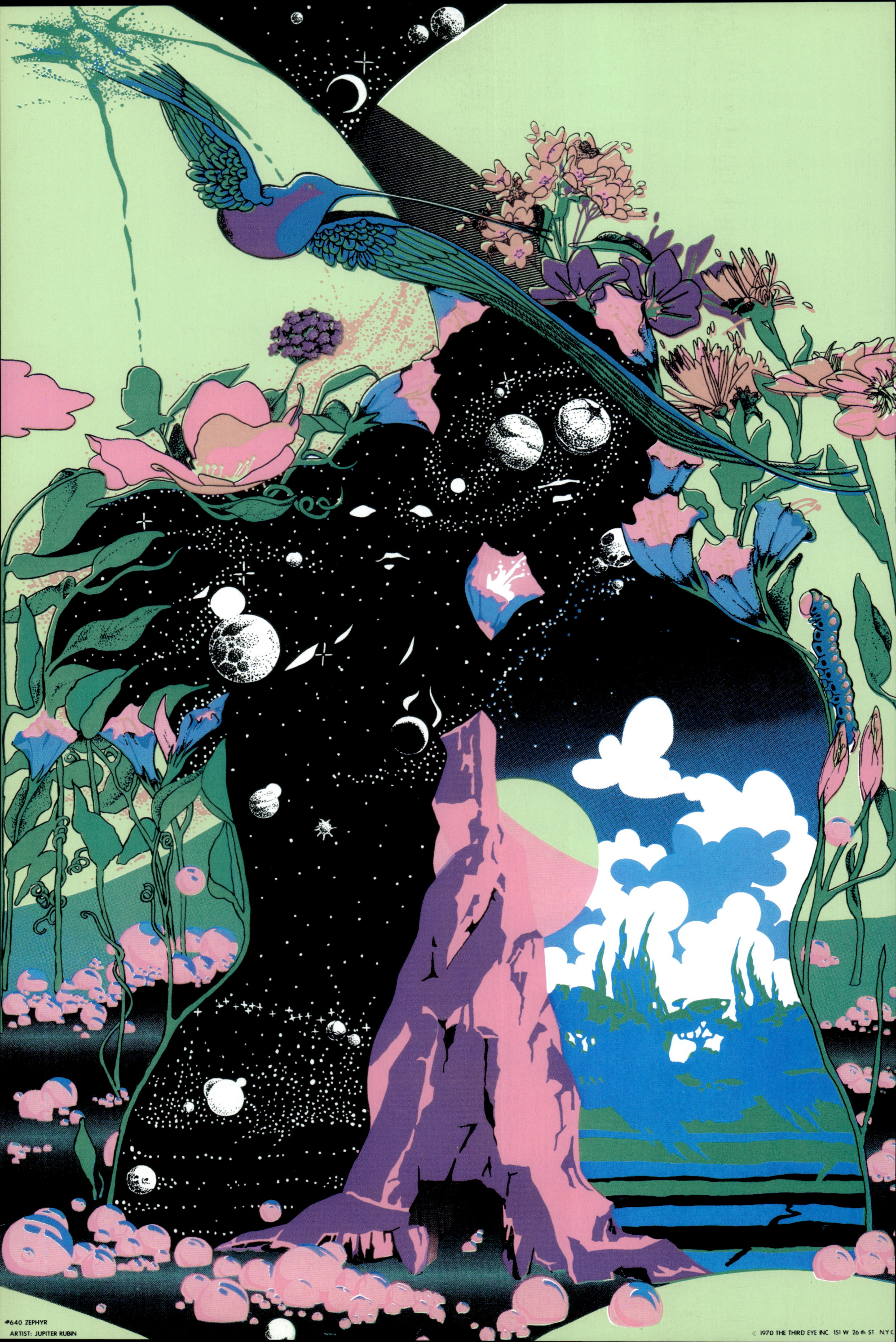
#640 ZEPHYR
ARTIST: JUPITER RUBIN
© 1970 THE THIRD EYE INC. 151 W 26th ST N.Y.C.

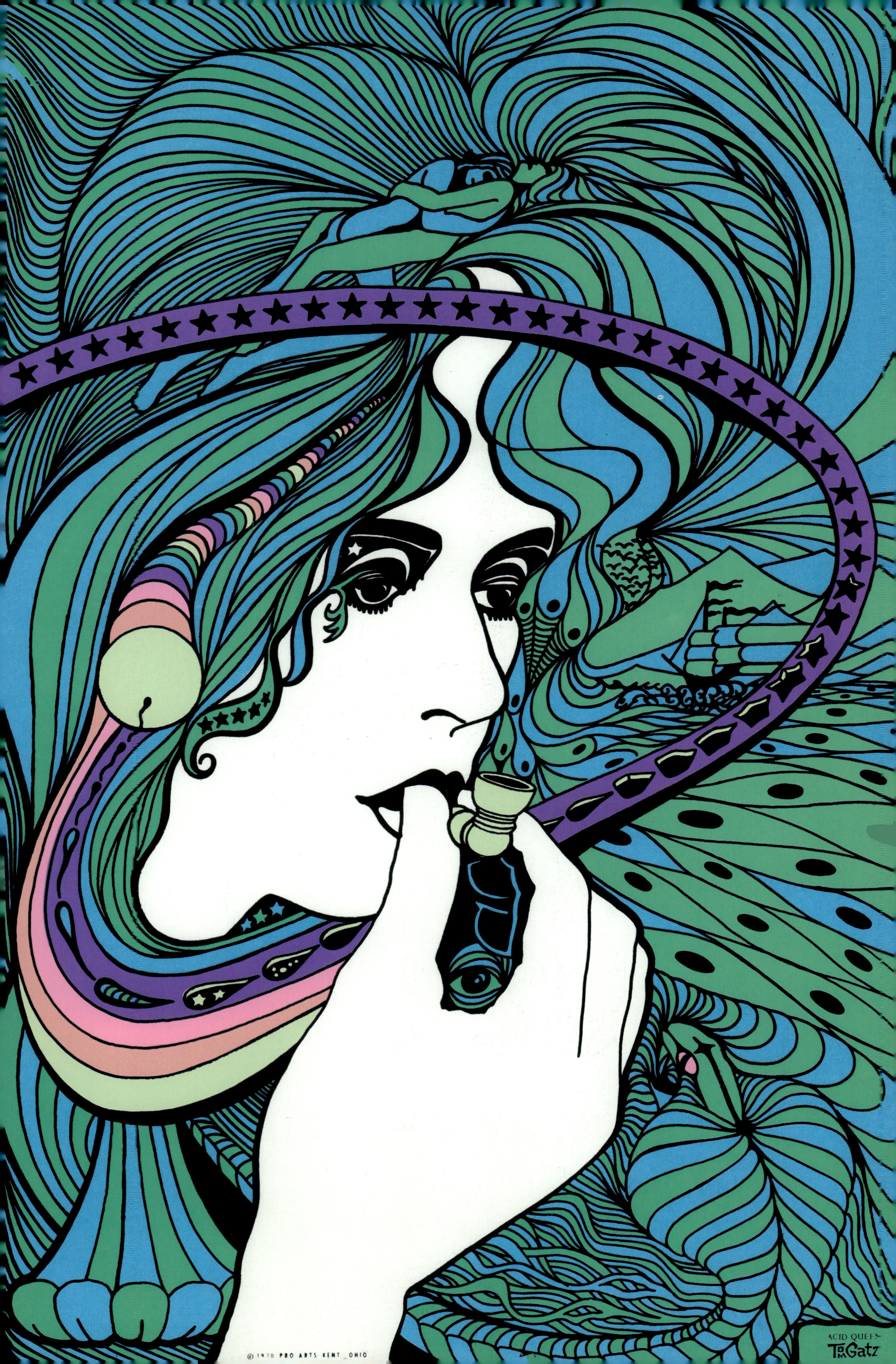
© 1970 PRO ARTS KENT OHIO
ACID QUEEN
TOM Gatz

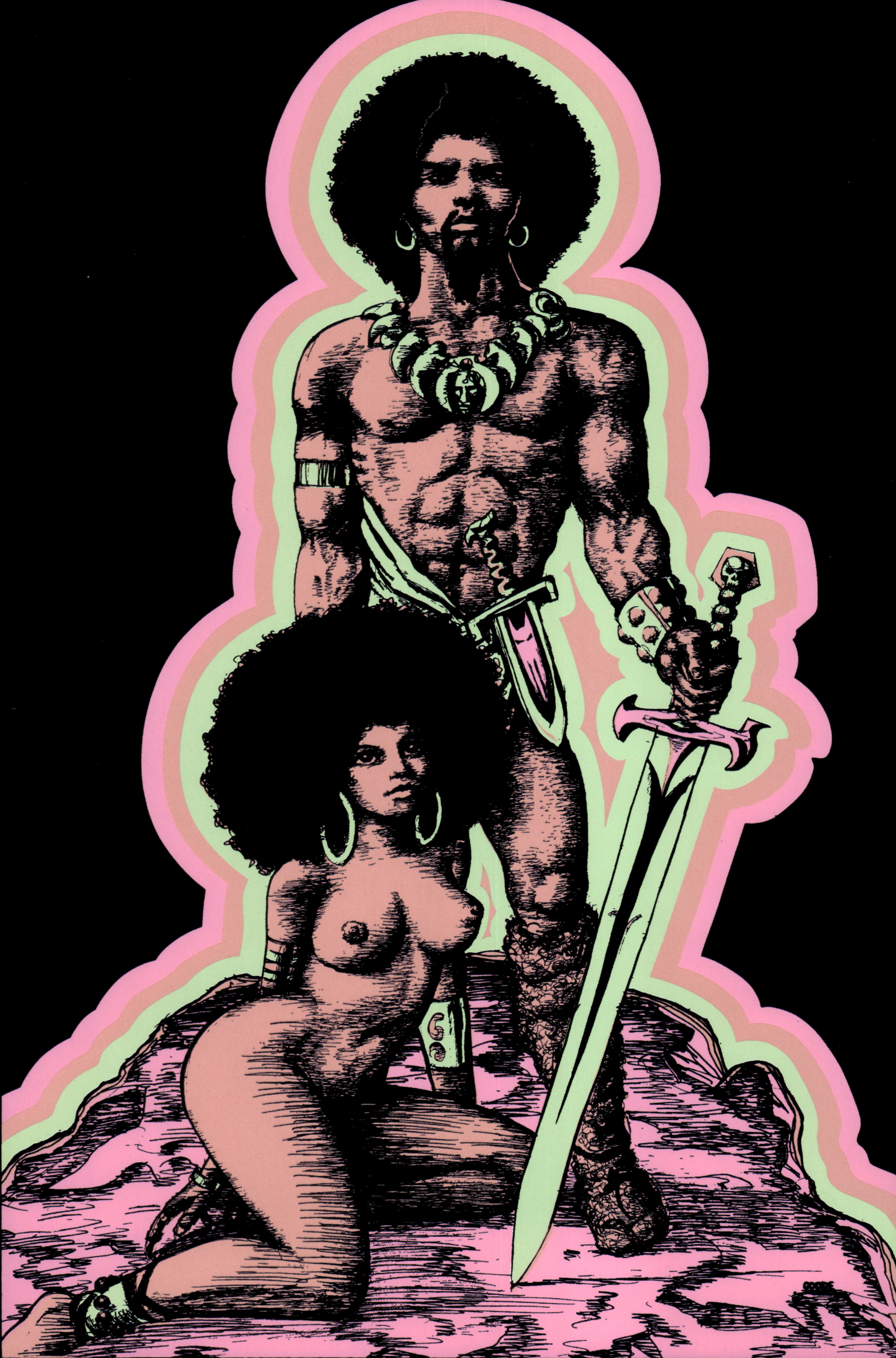

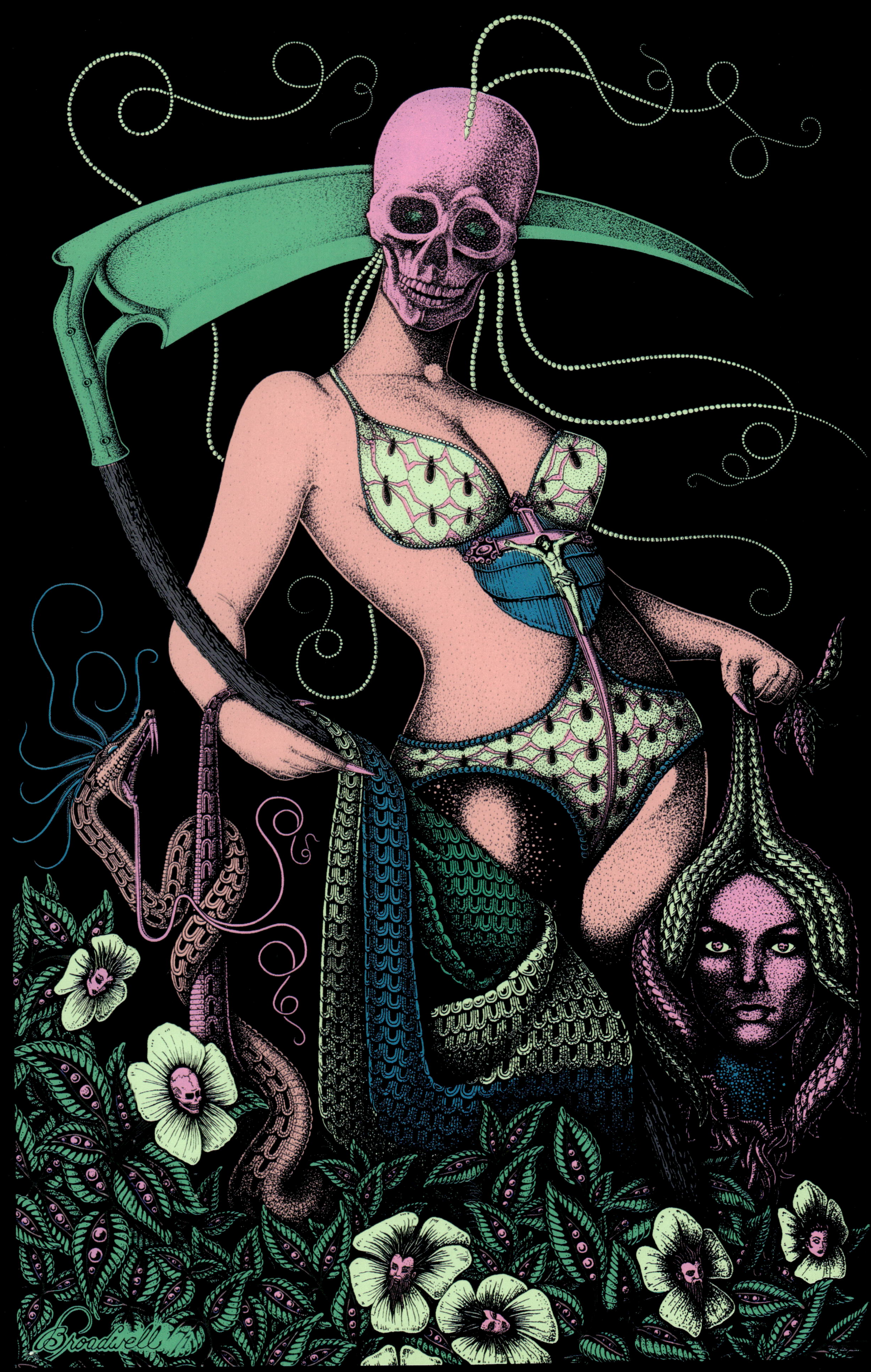
Broadwell

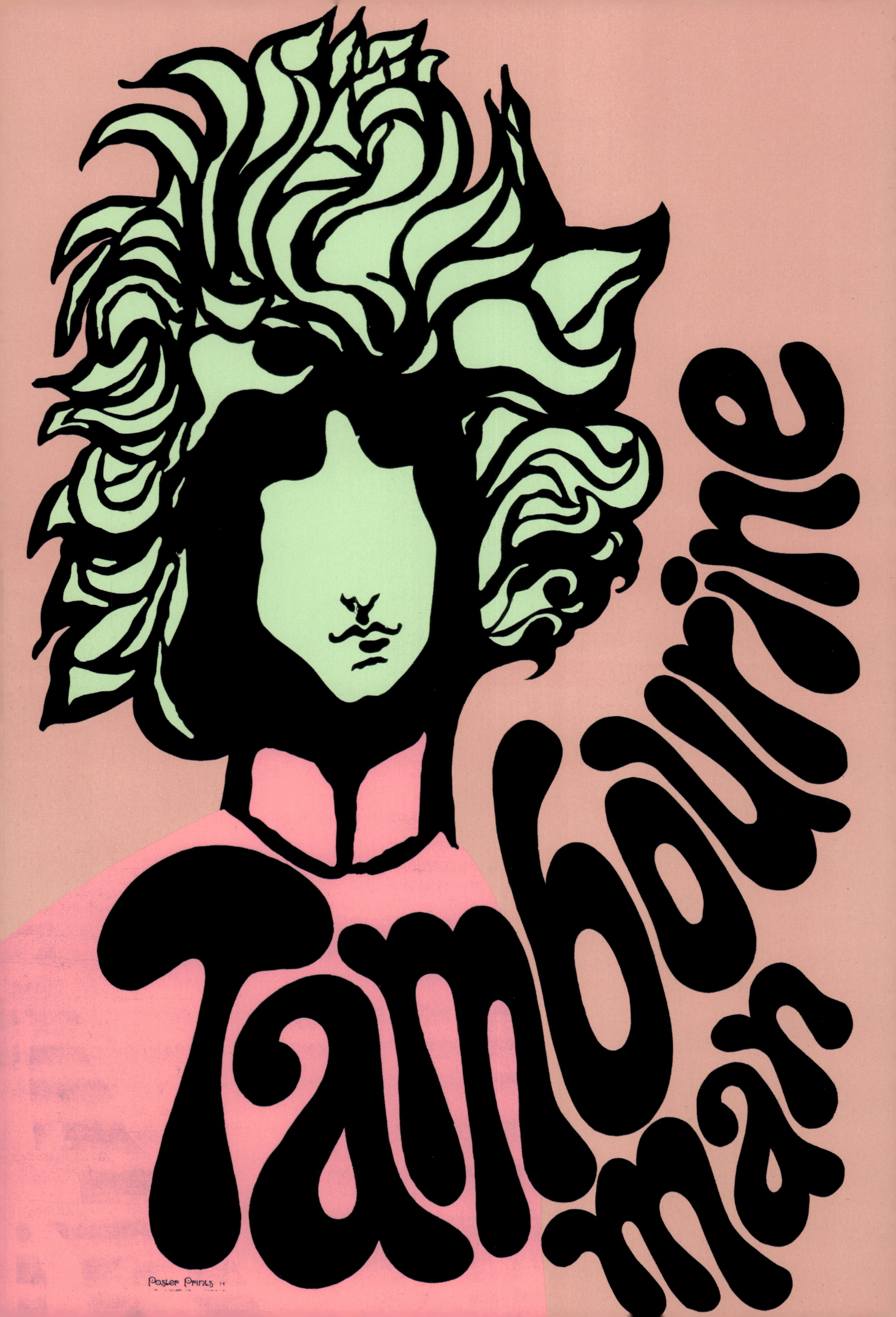
Tambourine
man
Poster Prints

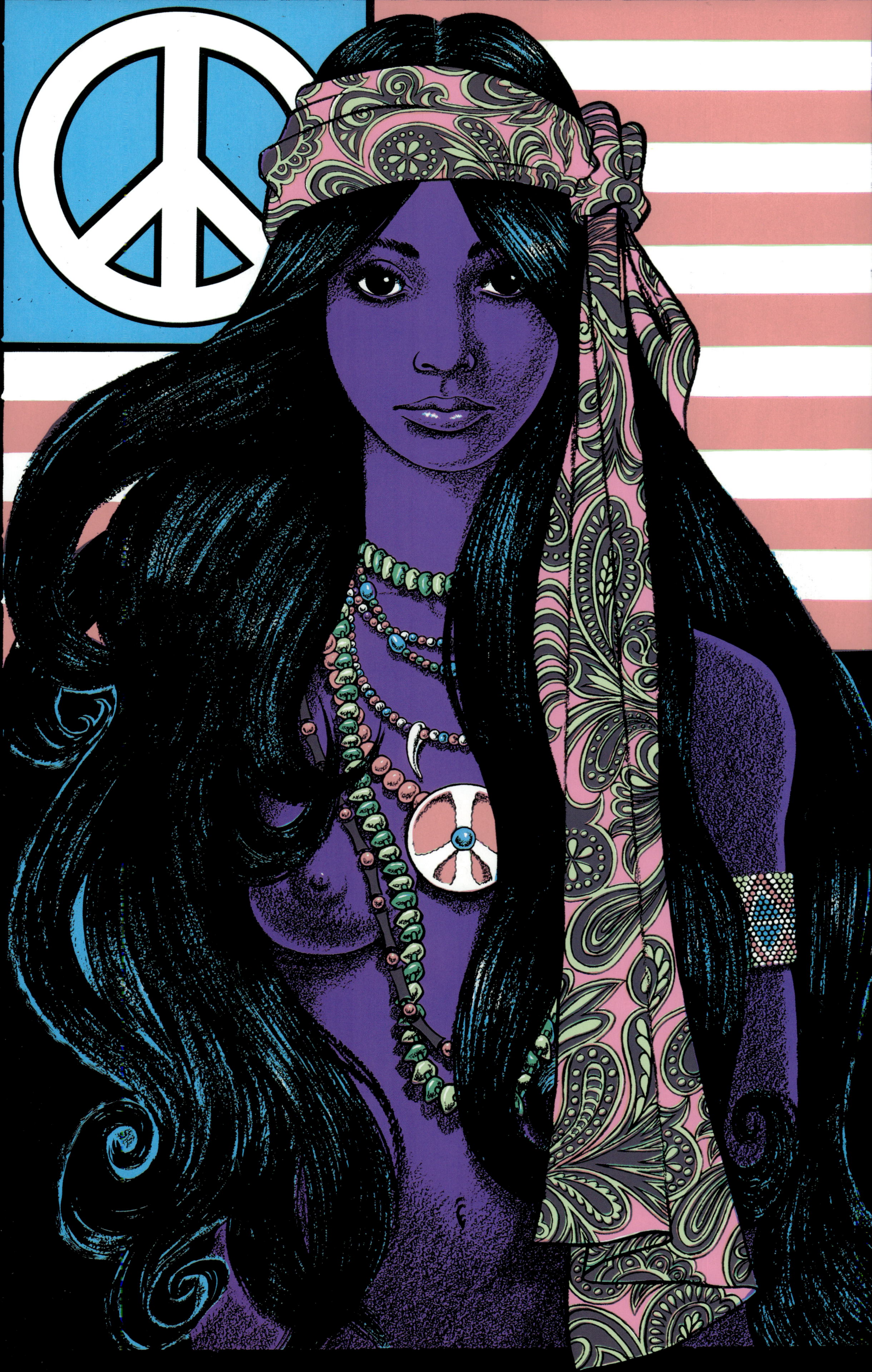

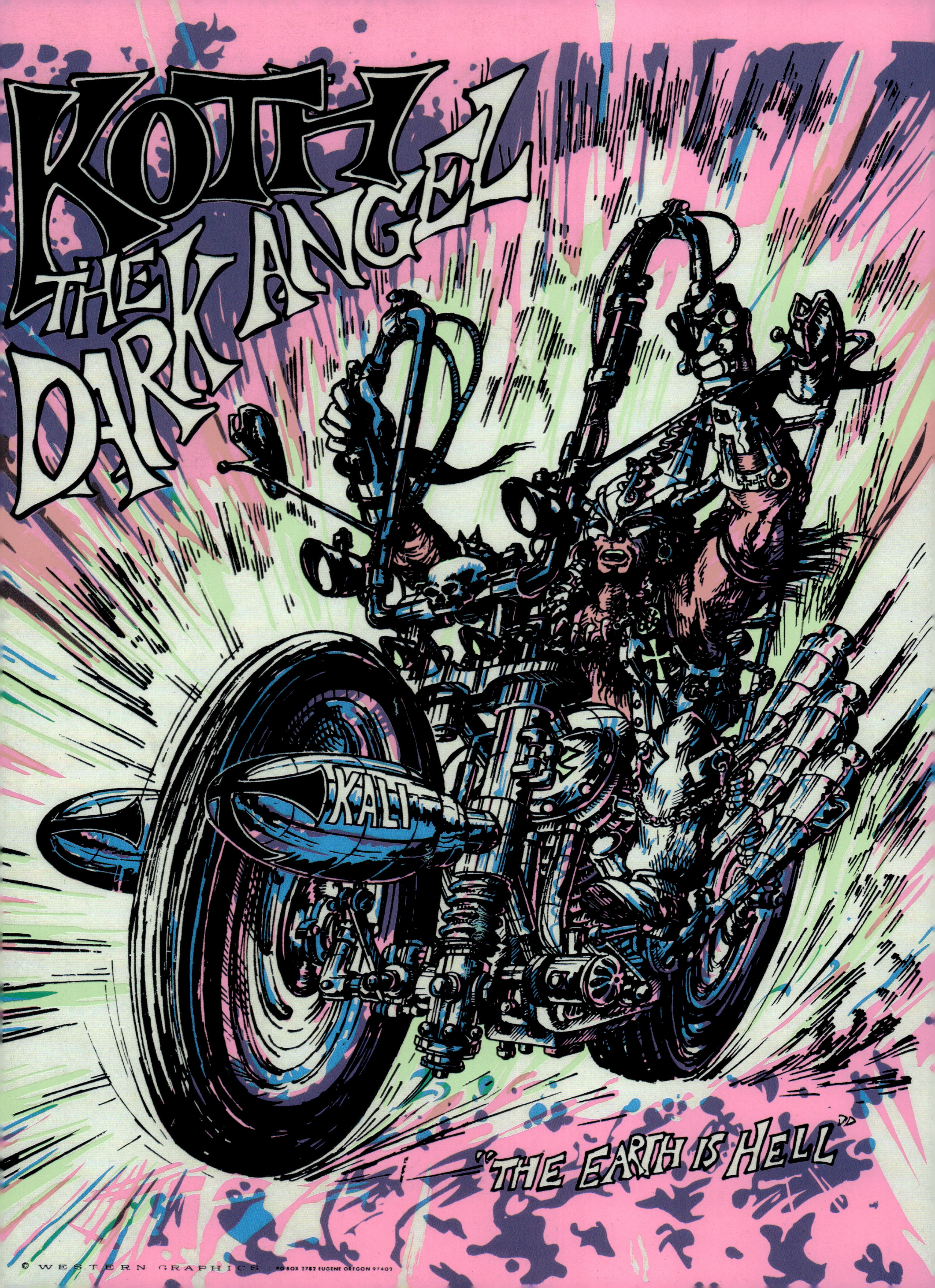
KOTH
THE DARK ANGEL
KALI
"THE EARTH IS HELL"
© WESTERN GRAPHICS PO BOX 2782 EUGENE OREGON 97402

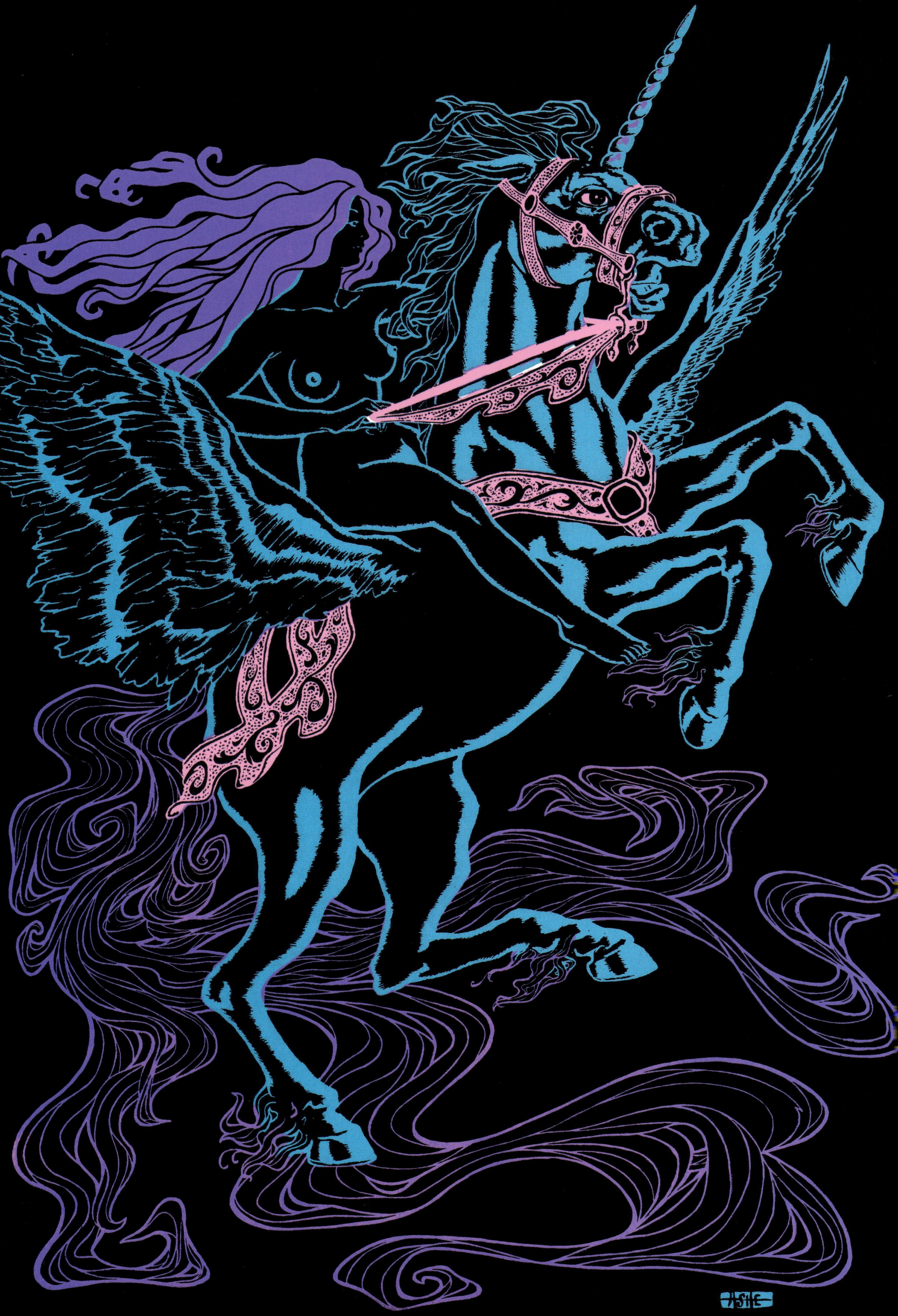

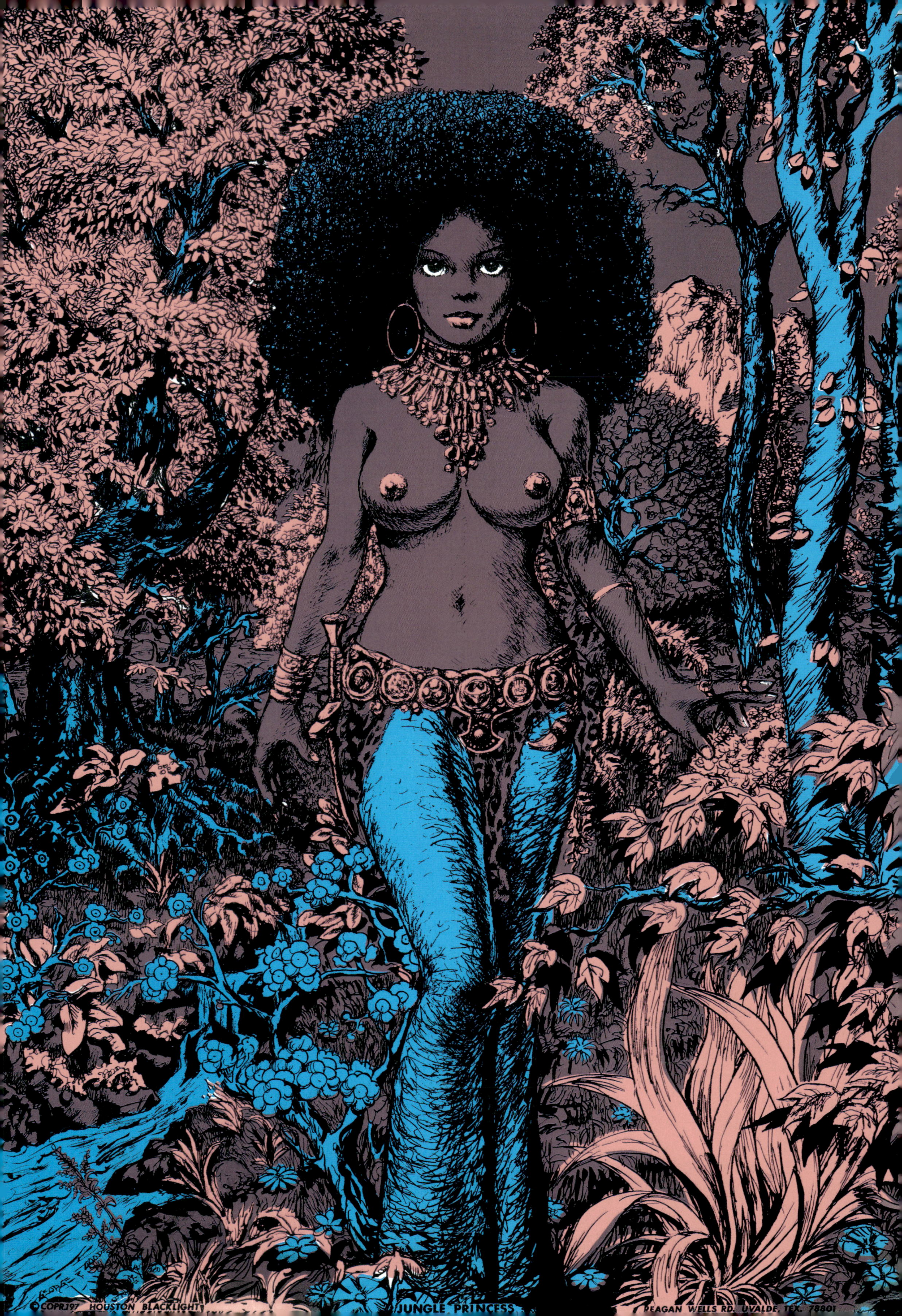
©COPR197_ HOUSTON BLACKLIGHT
JUNGLE PRINCESS
REAGAN WELLS RD. UVALDE, TEX. 78801

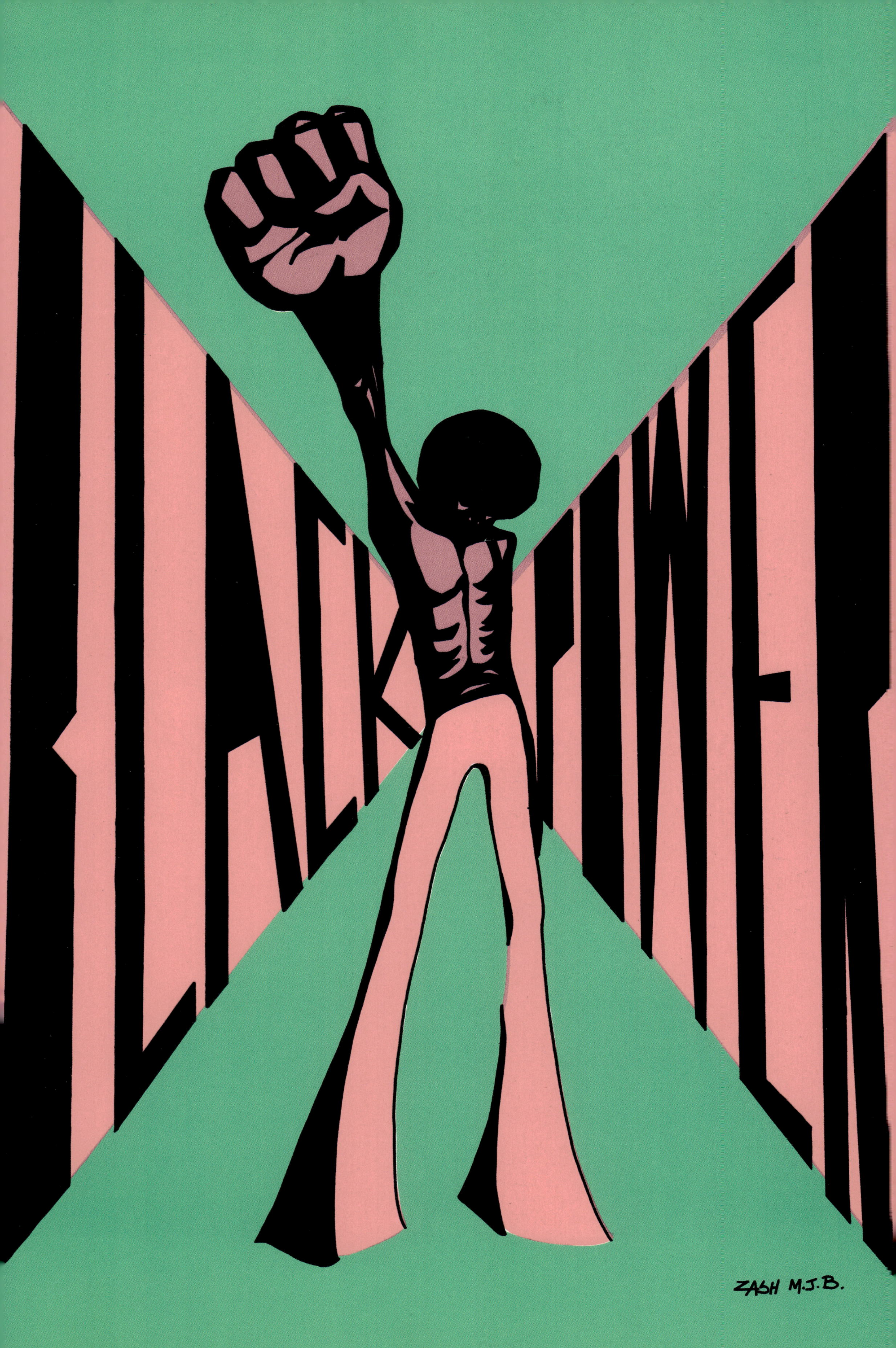
BLACK POWER
ZASH M.J.B.

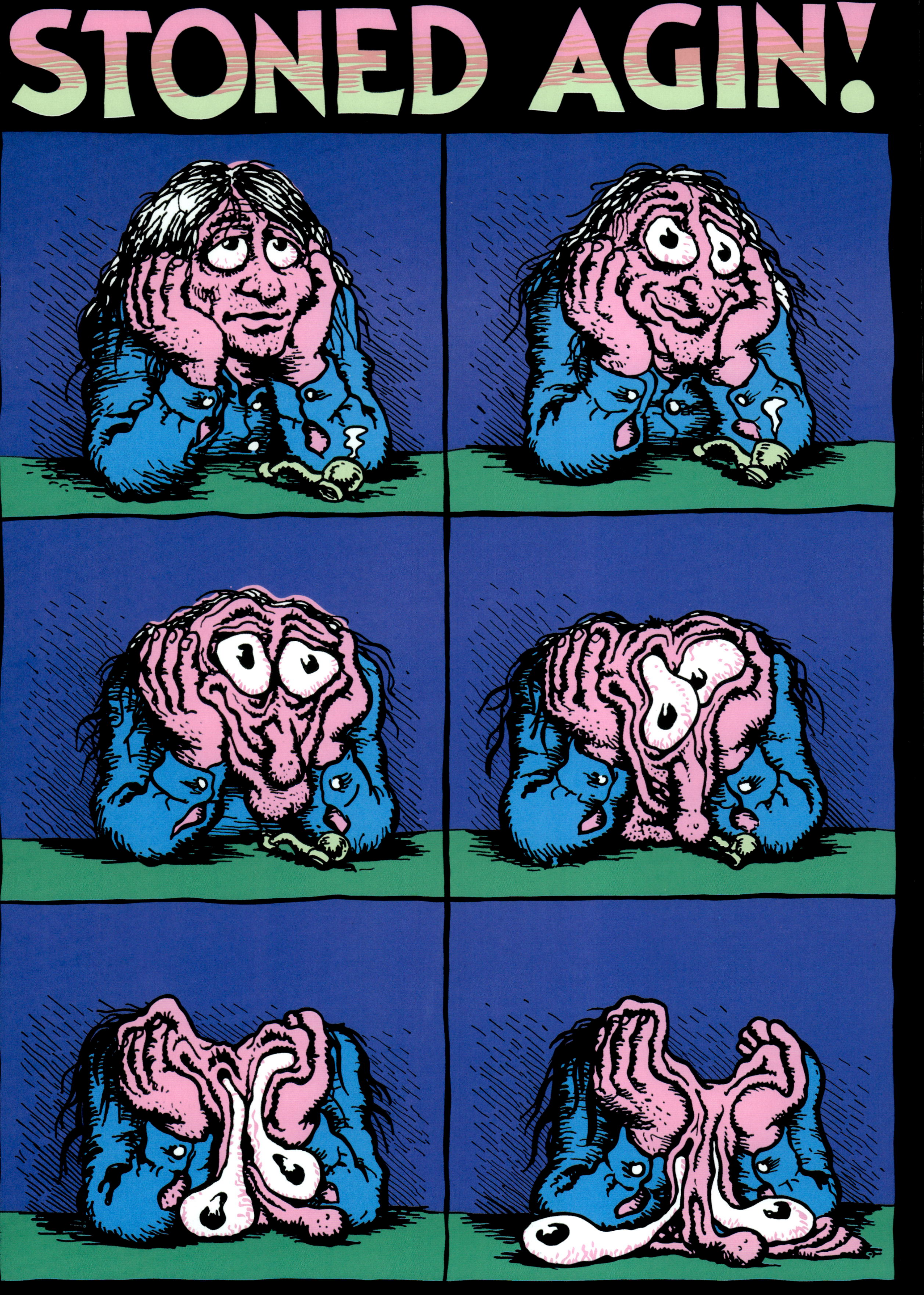
STONED AGIN!

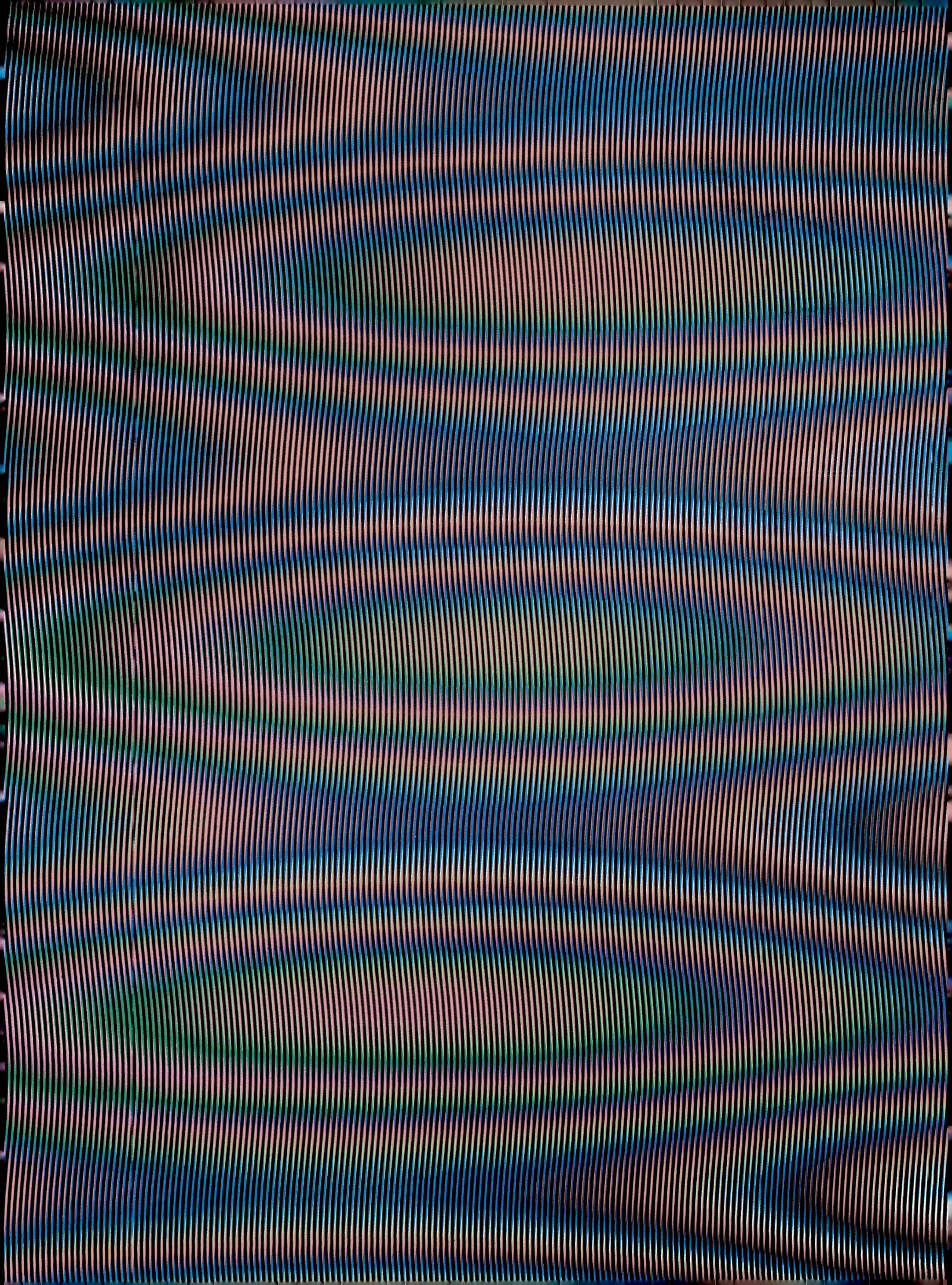

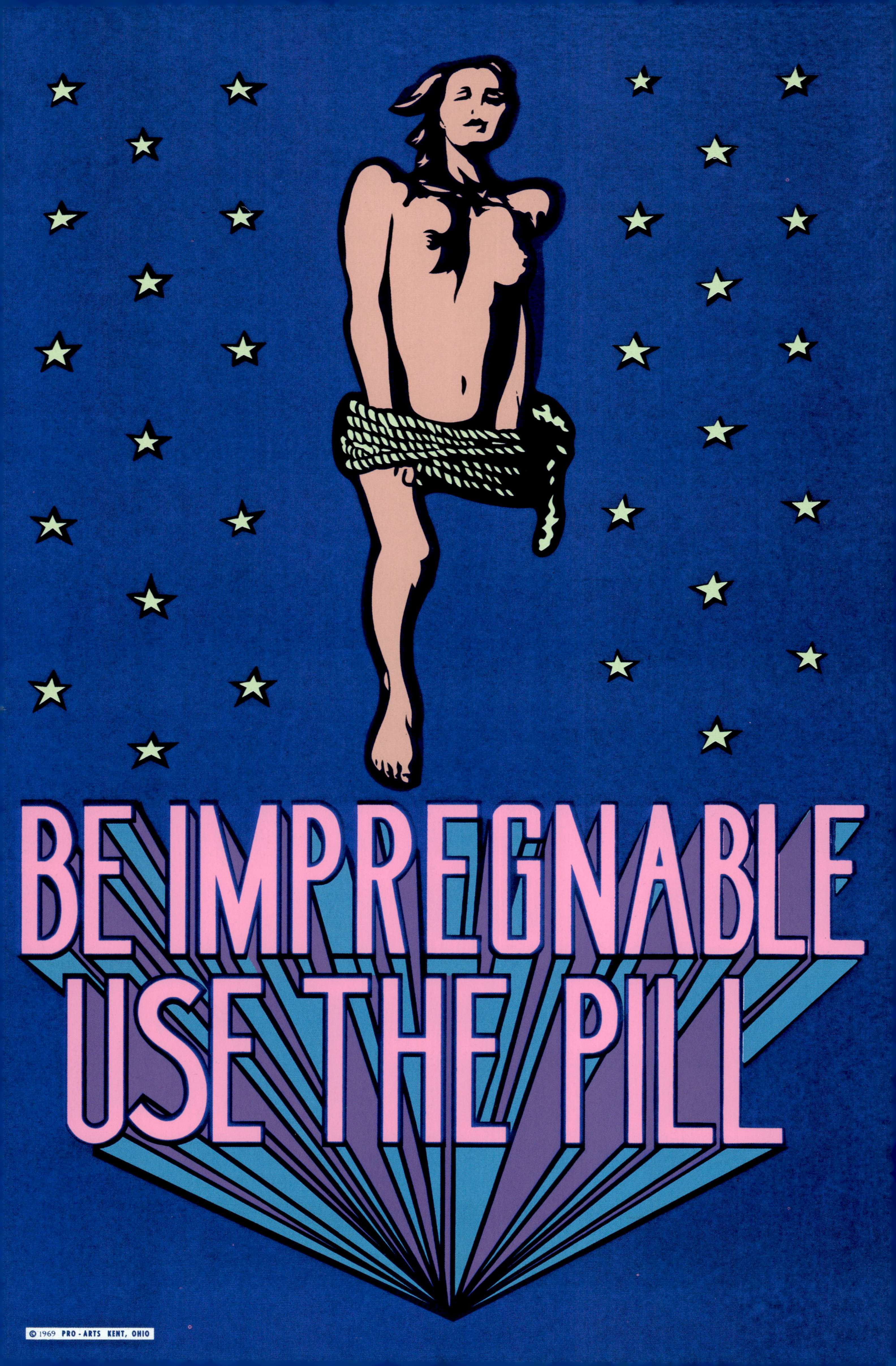
BE IMPREGNABLE
USE THE PILL
© 1969 PRO-ARTS KENT, OHIO

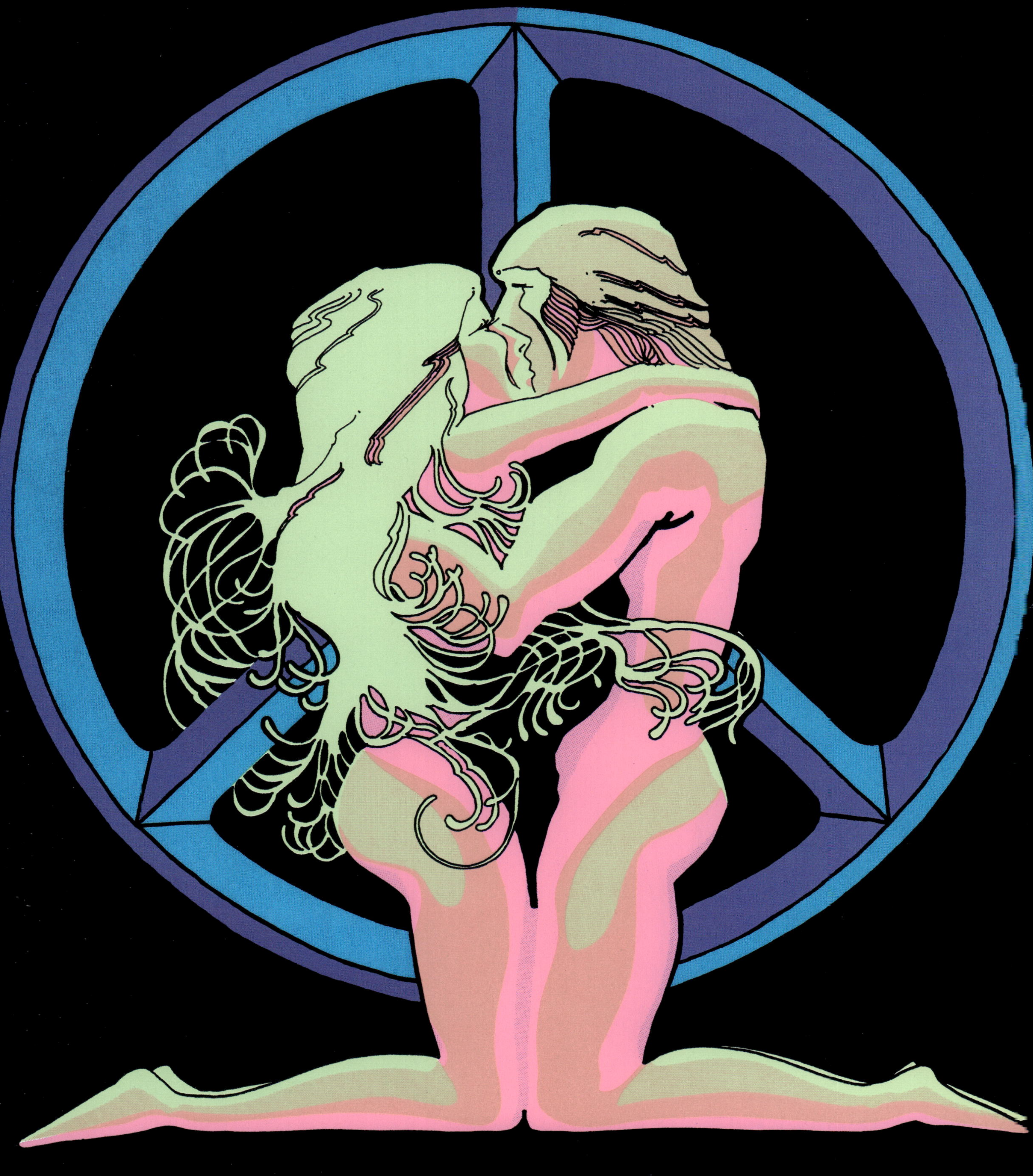

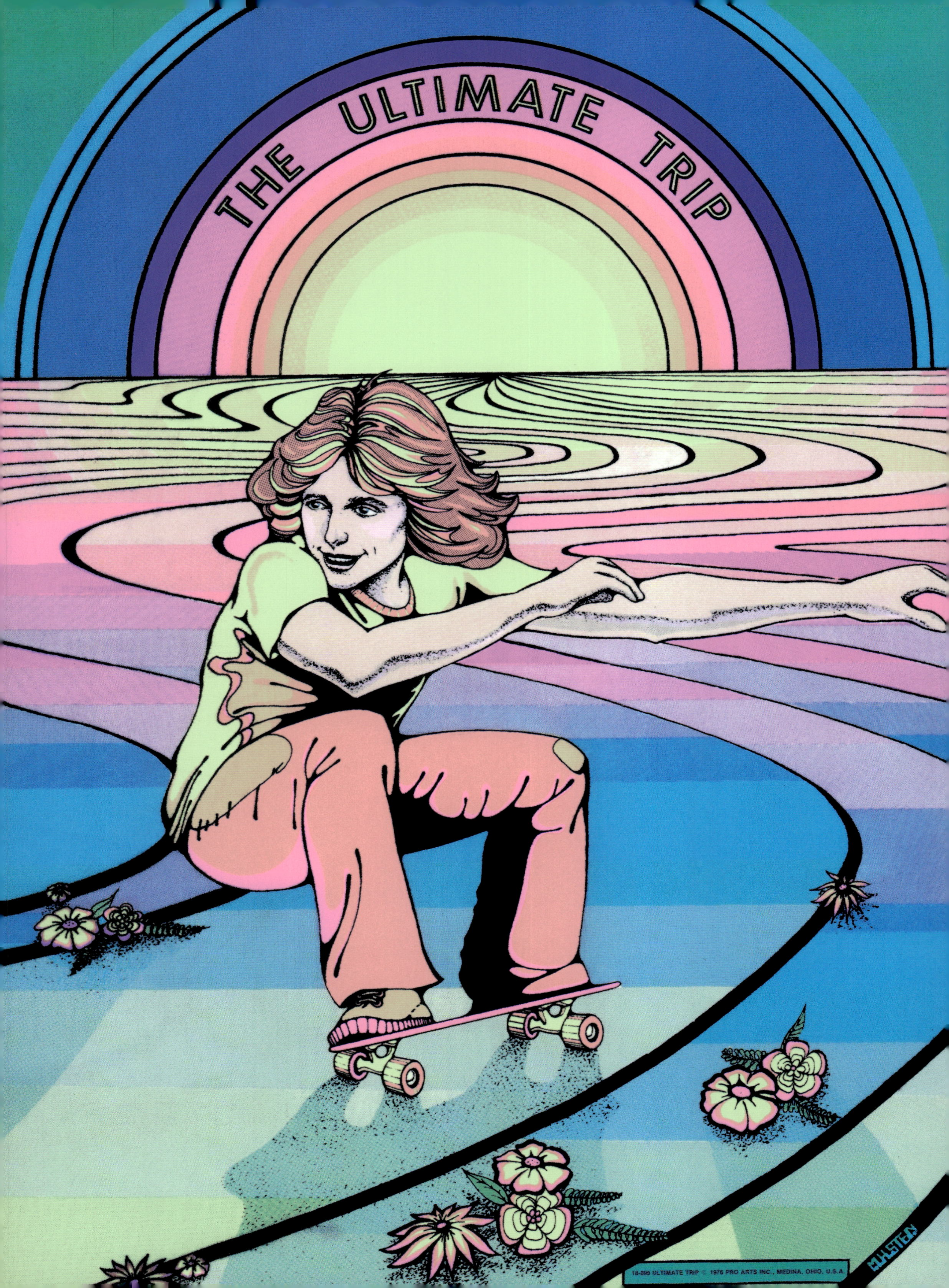
THE ULTIMATE TRIP
18-899 ULTIMATE TRIP © 1976 PRO ARTS INC., MEDINA, OHIO, U.S.A.

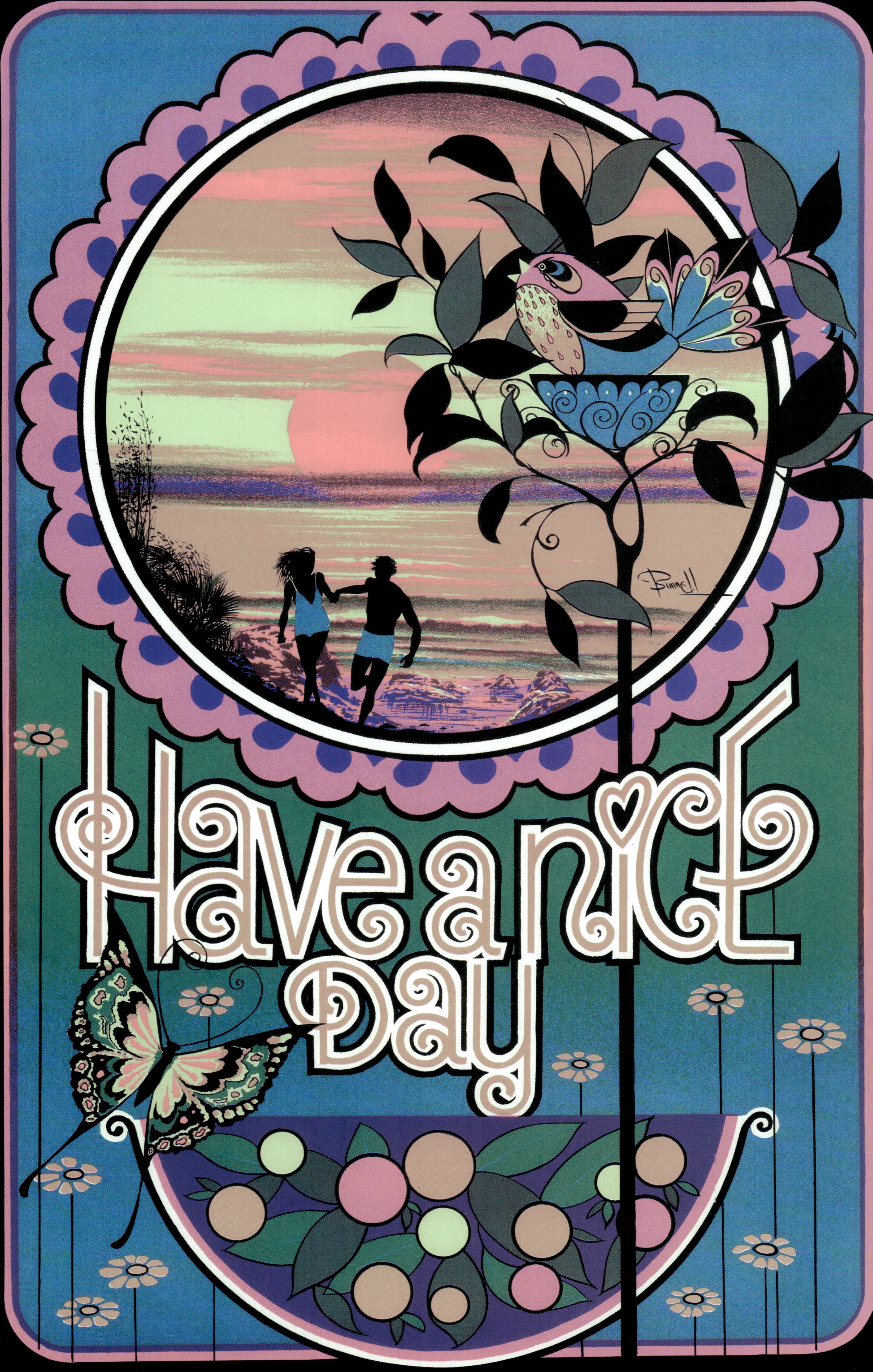
Have a nice Day
Bunnell

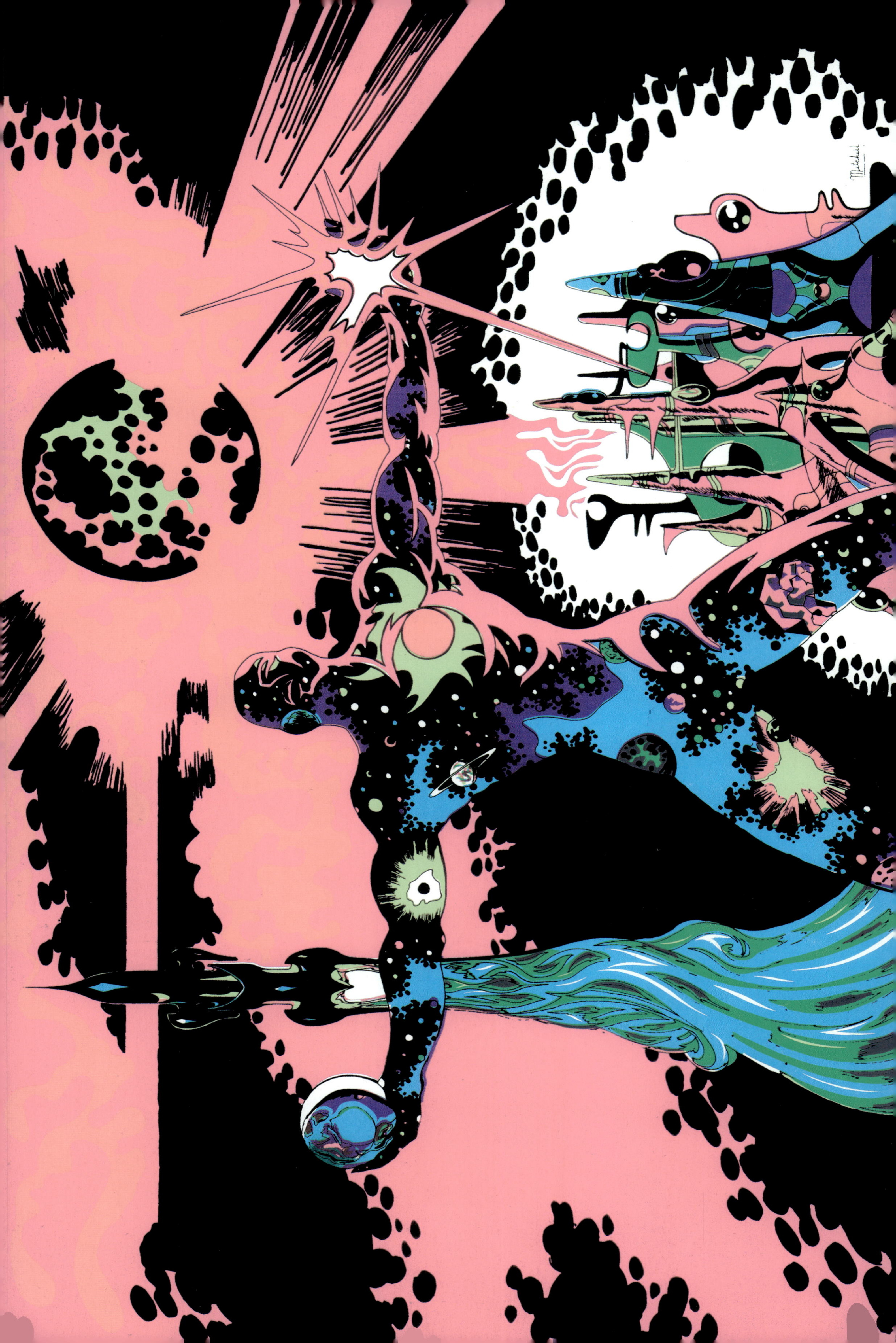

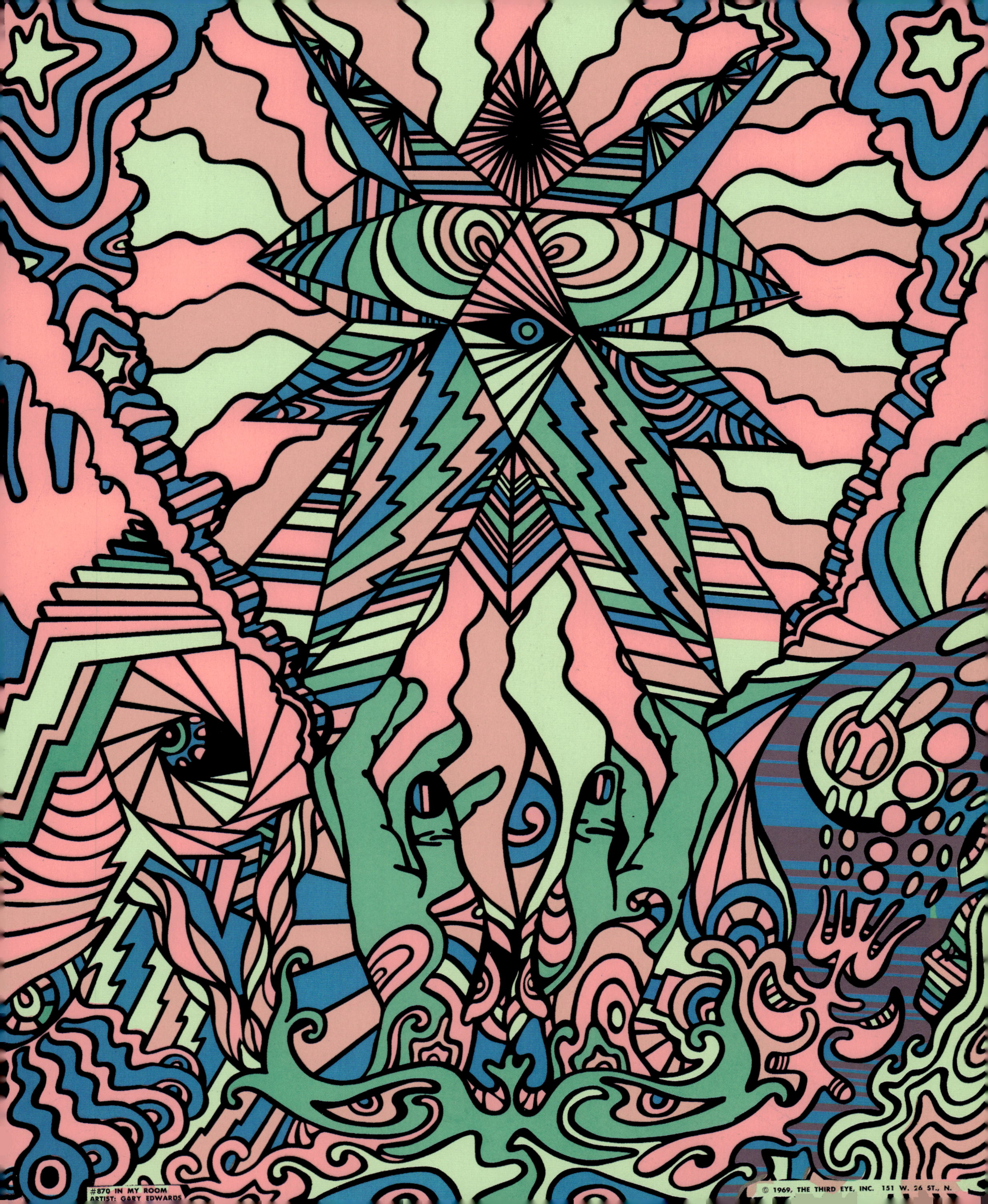
#870 IN MY ROOM
ARTIST: GARY EDWARDS
© 1969, THE THIRD EYE, INC. 151 W. 26 ST., N.

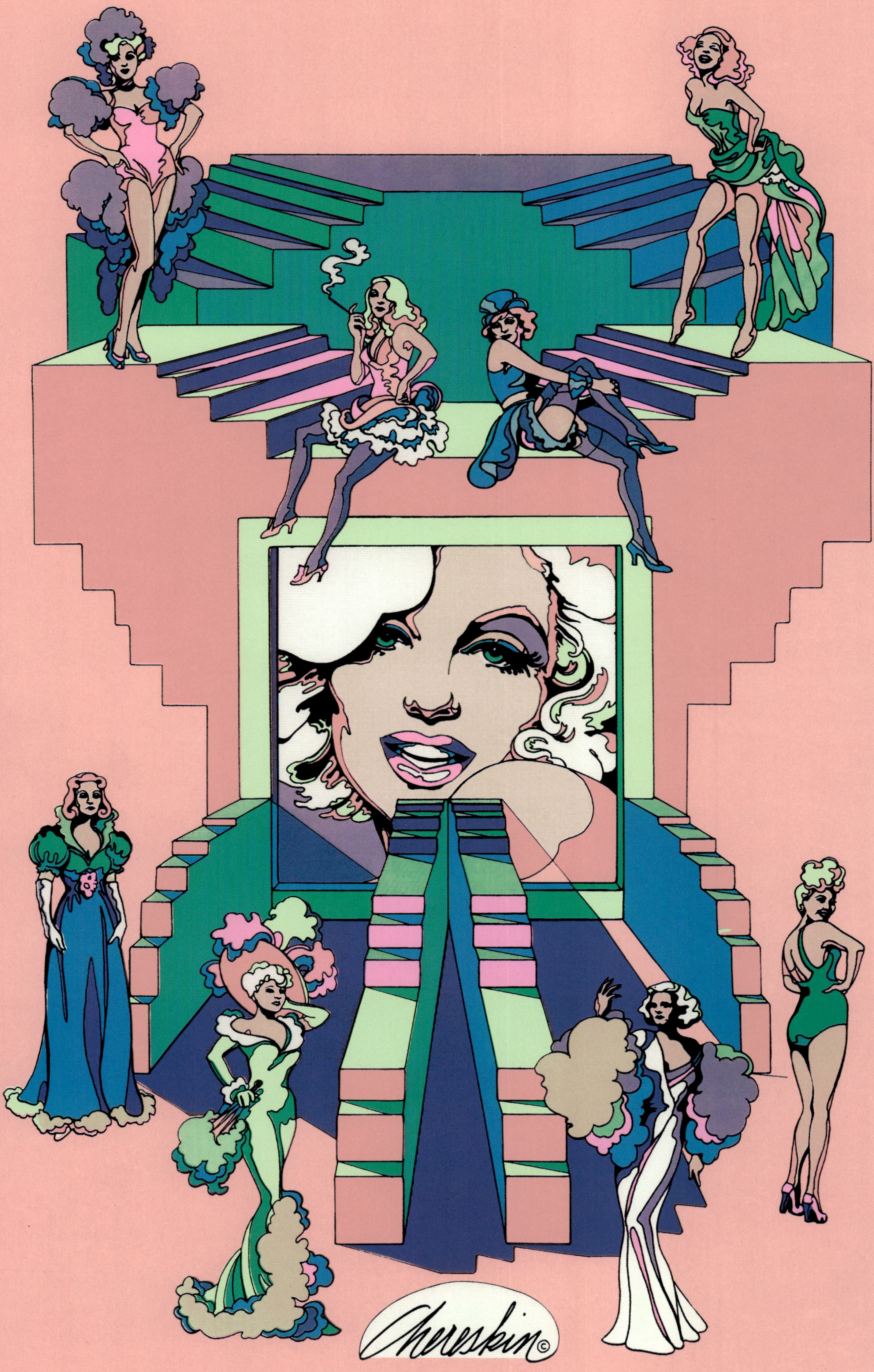

ARTKO STUDIO IMS CORPORATION BOX 3399 ALBUQUERQUE, NEW MEXICO 87110 PRINTED IN U.S.A. MOVIE QUEENS R522

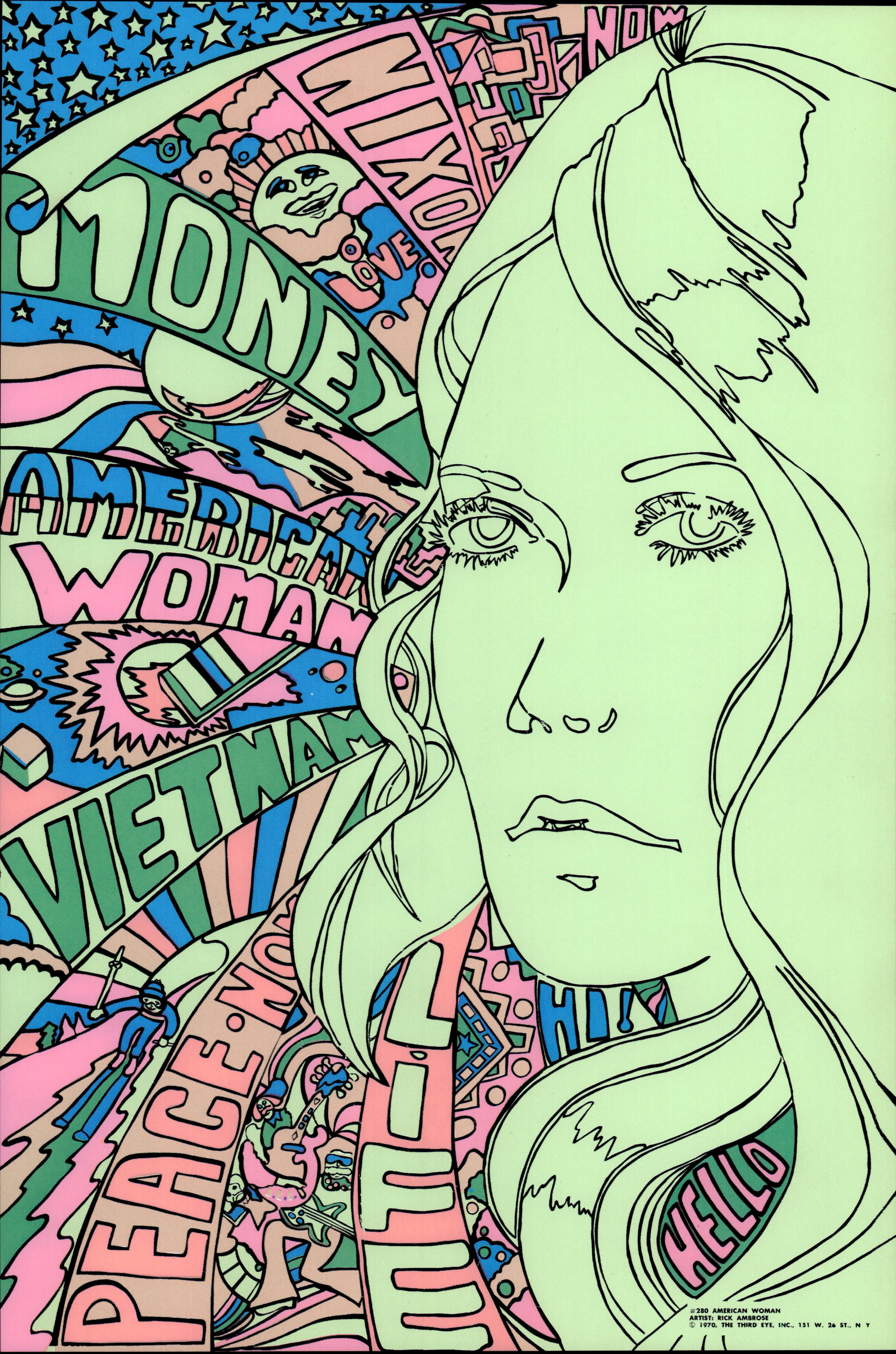
NOW
NIXON
MONEY
LOVE
AMERICAN
WOMAN
VIETNAM
PEACE-NOW
HI!
HELLO
#280 AMERICAN WOMAN
ARTIST: RICK AMBROSE
© 1970, THE THIRD EYE, INC., 151 W. 26 ST., N Y

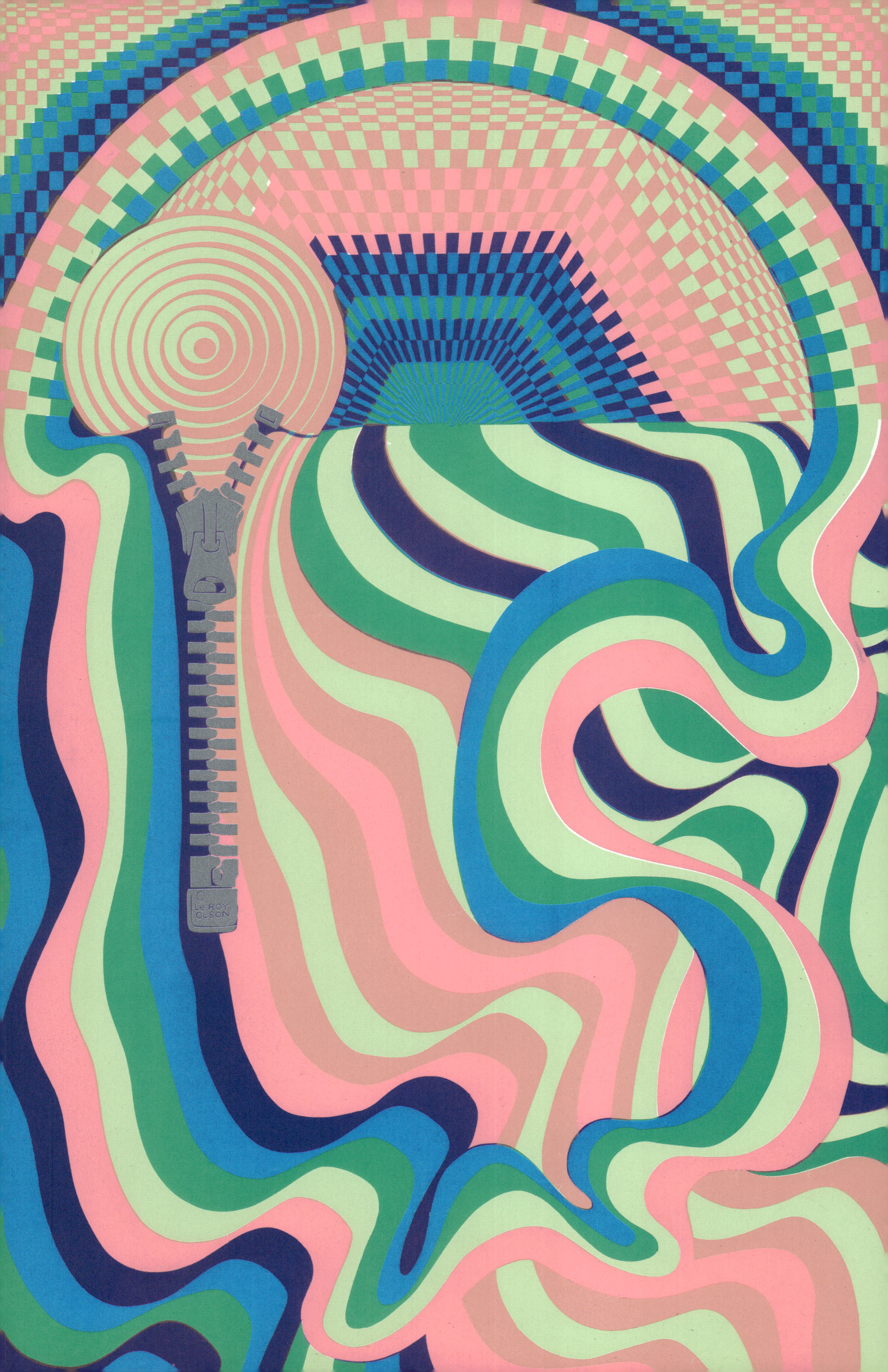
LeROY OLSON

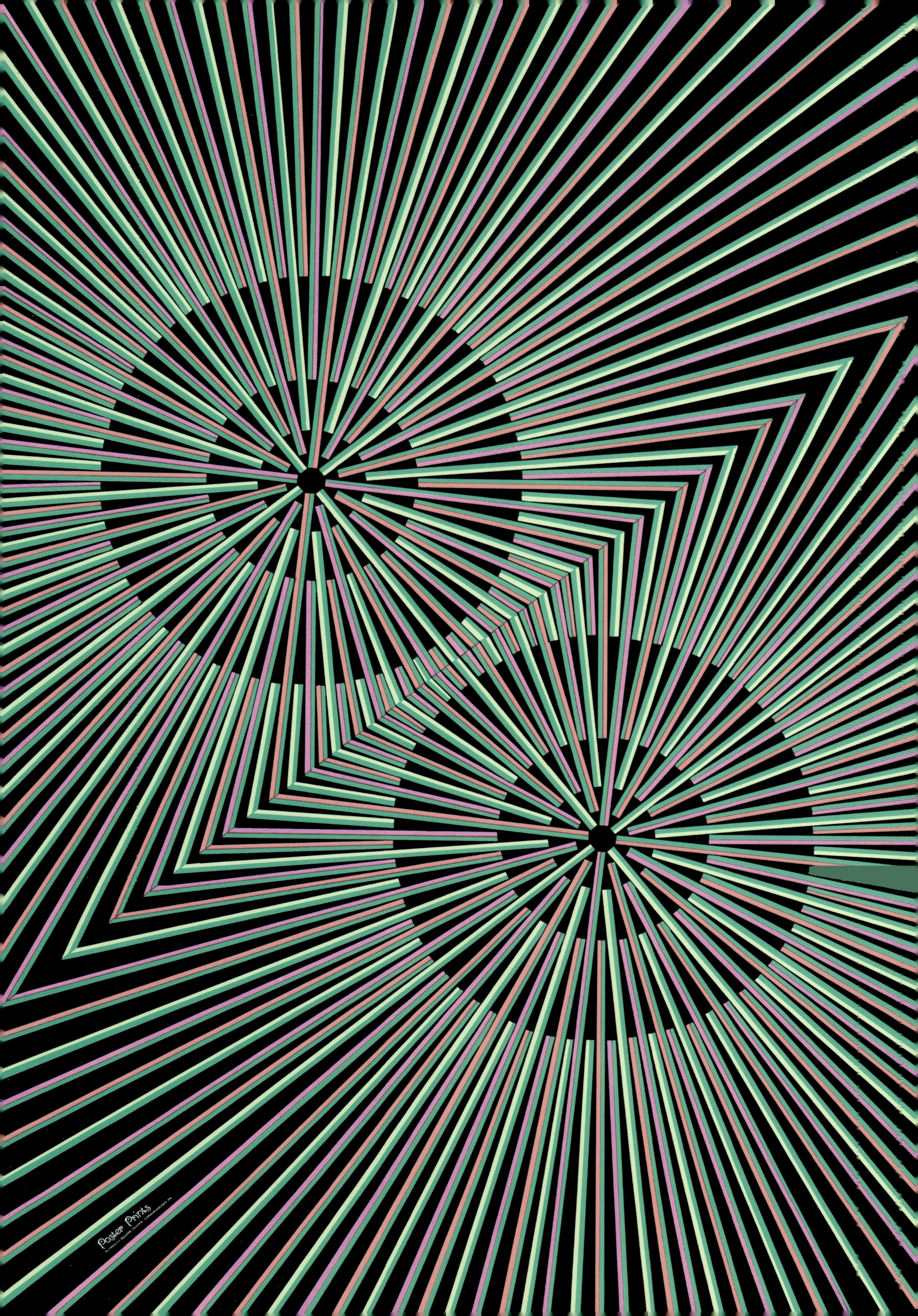
Poster Prints
PLYMOUTH SQUARE CENTER · CONSHOHOCKEN, PA

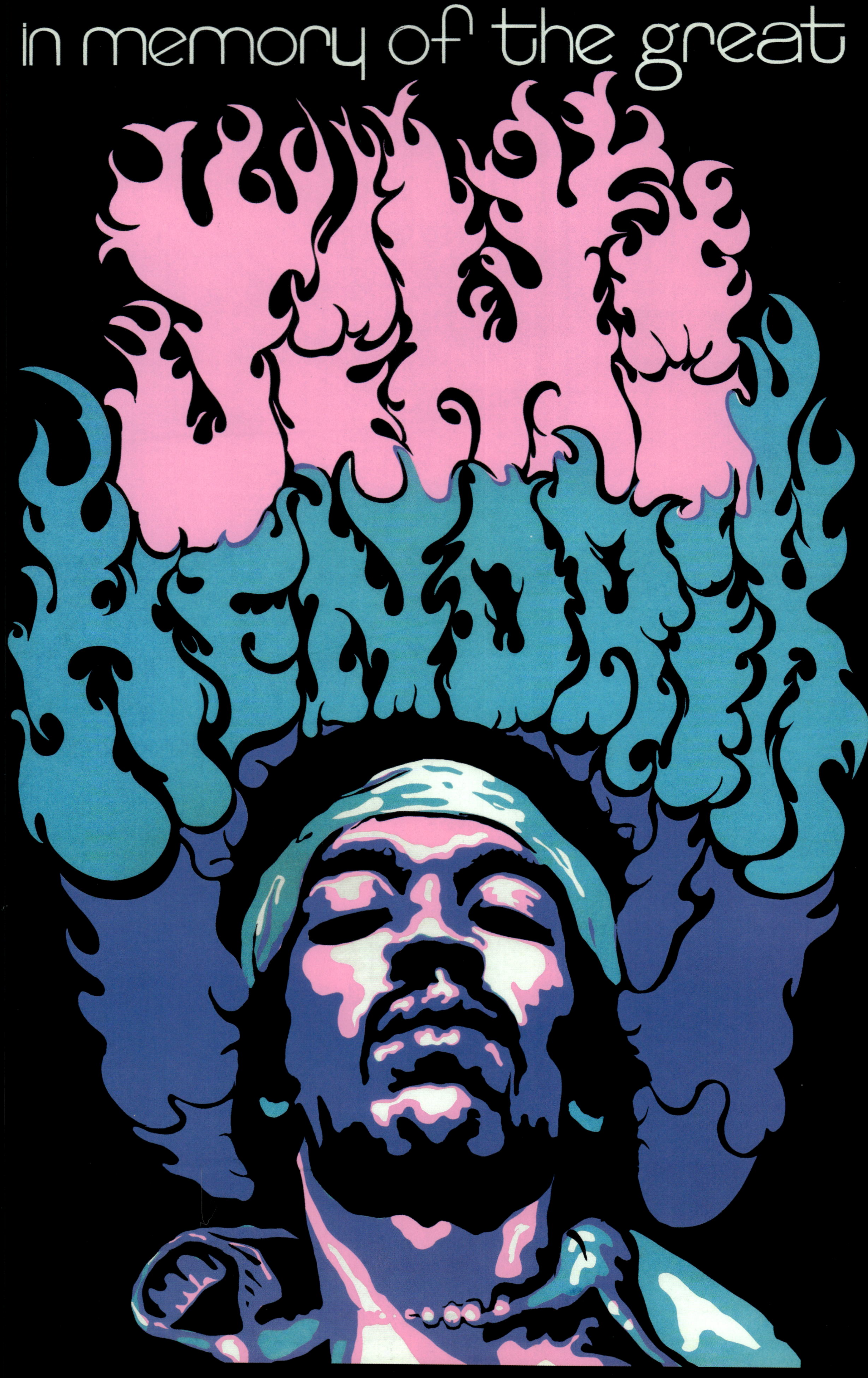
in memory of the great

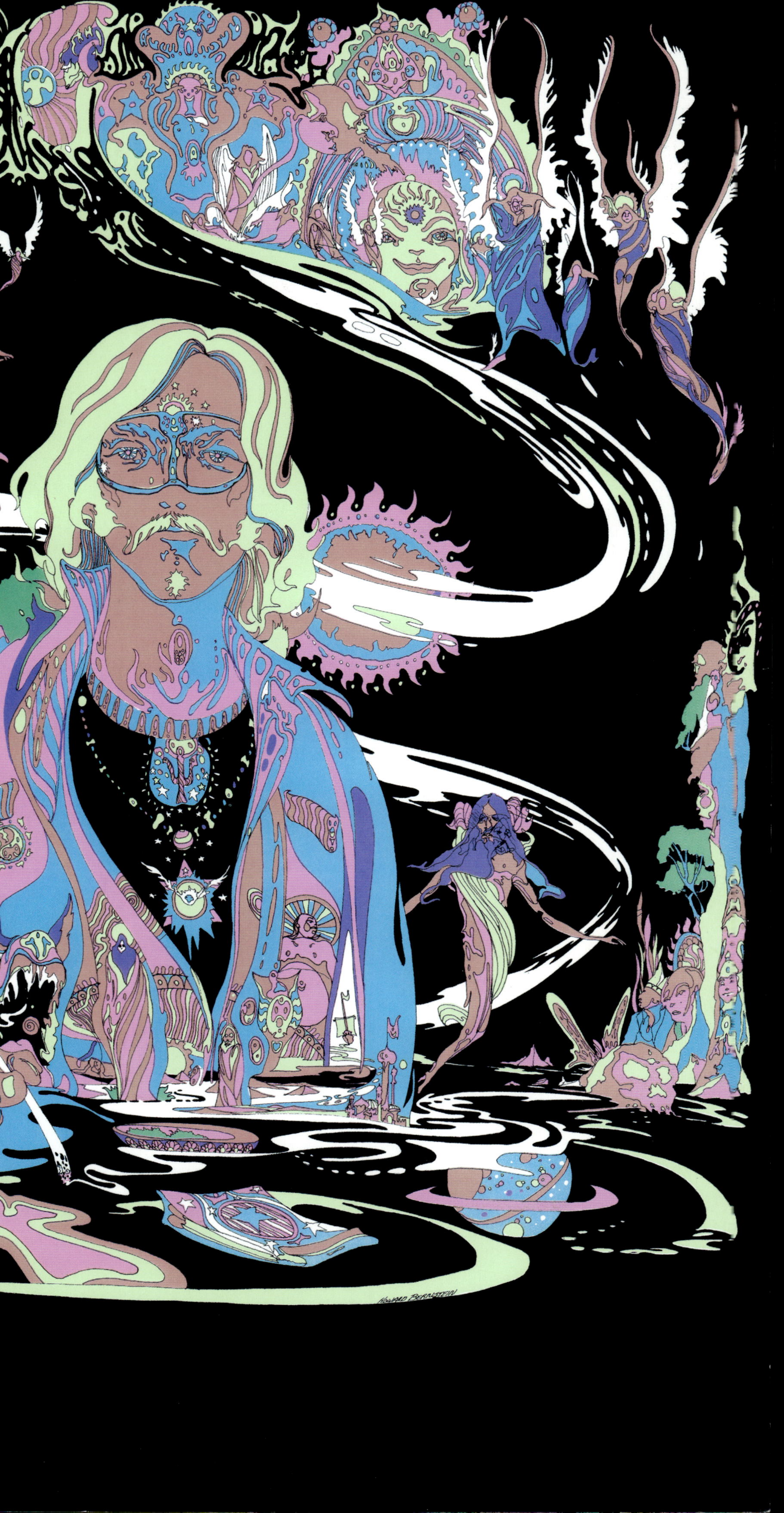
HOWARD BERNSTEIN

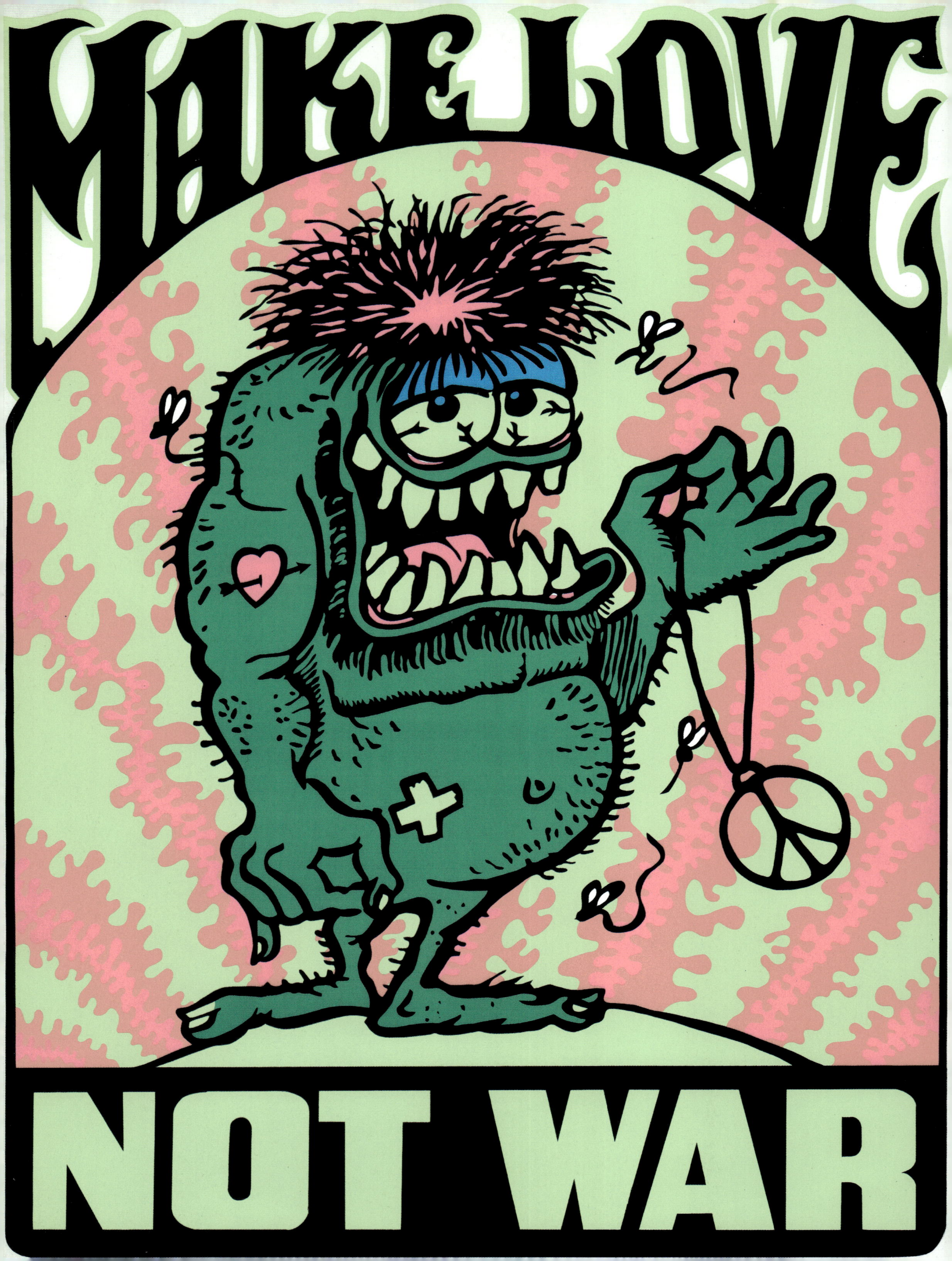
MAKE LOVE
NOT WAR

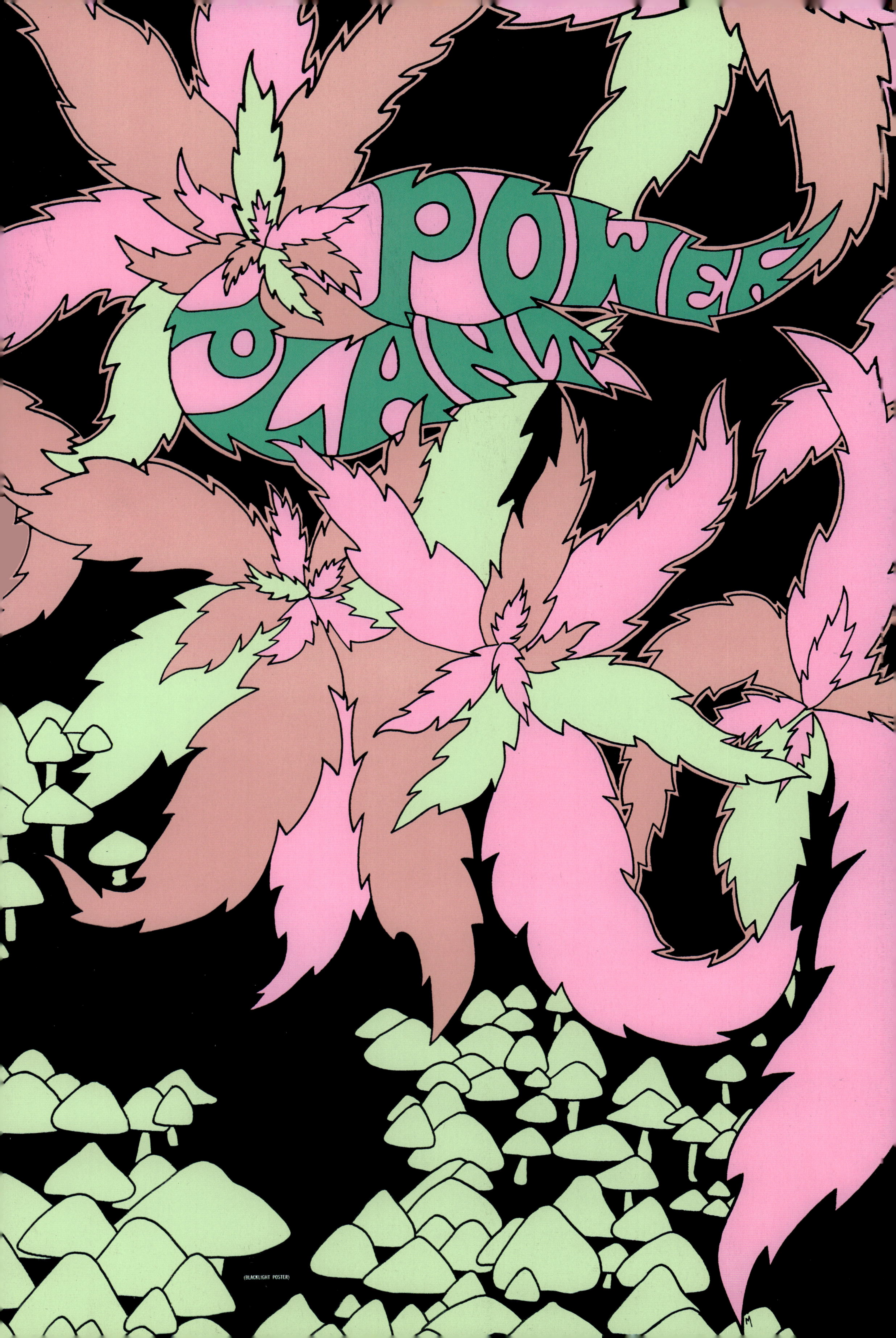
POWER
PLANT
(BLACKLIGHT POSTER)
M

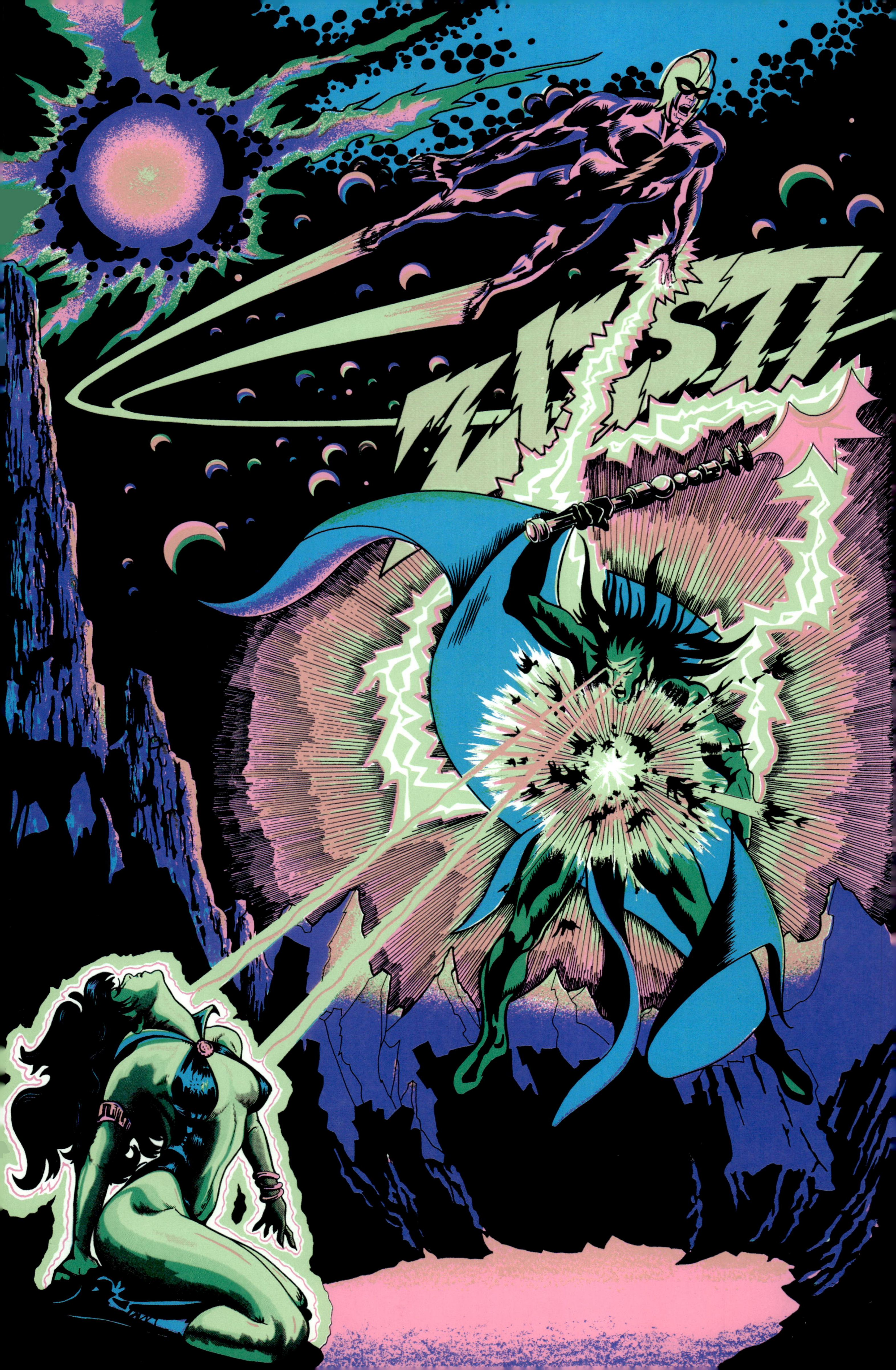

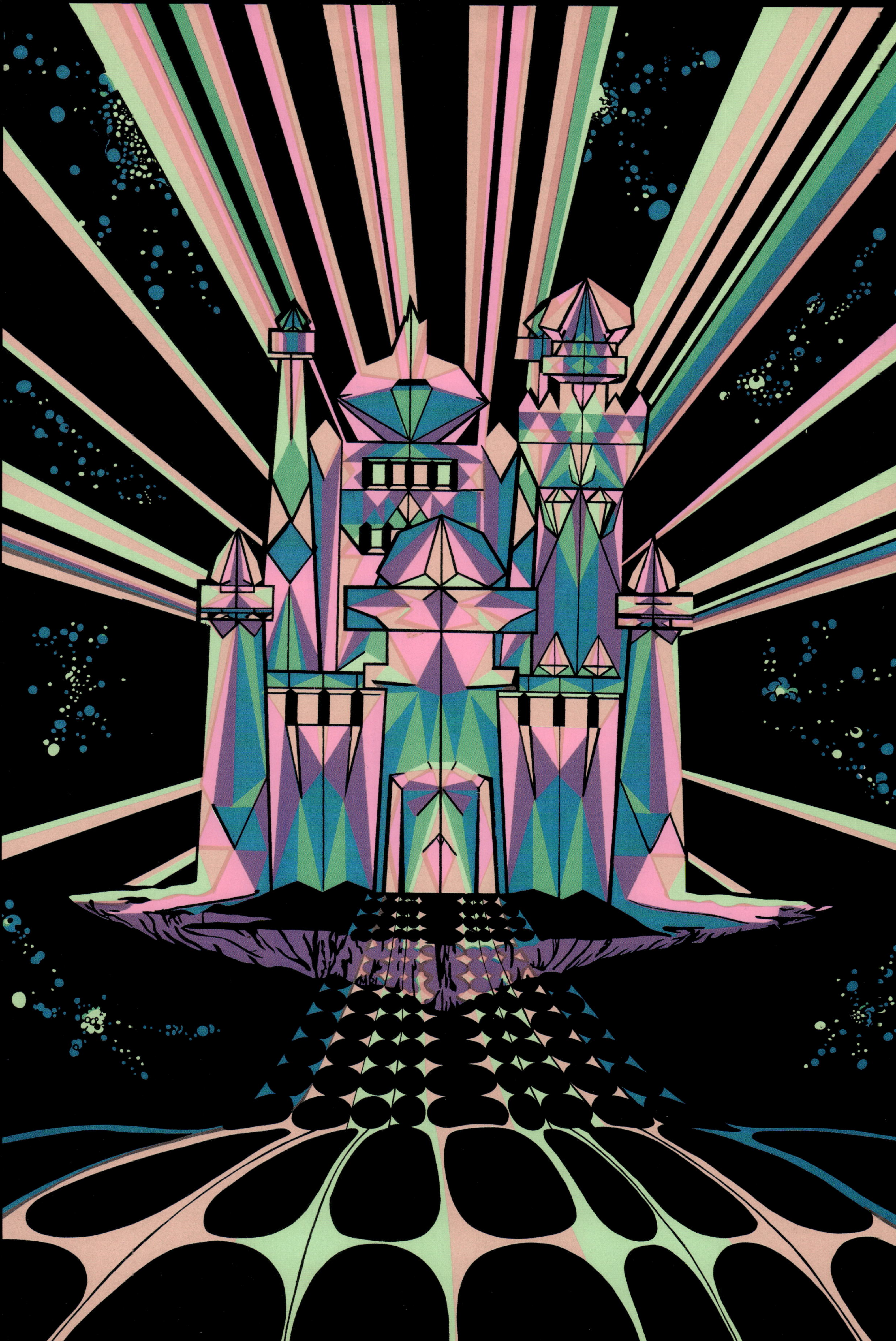

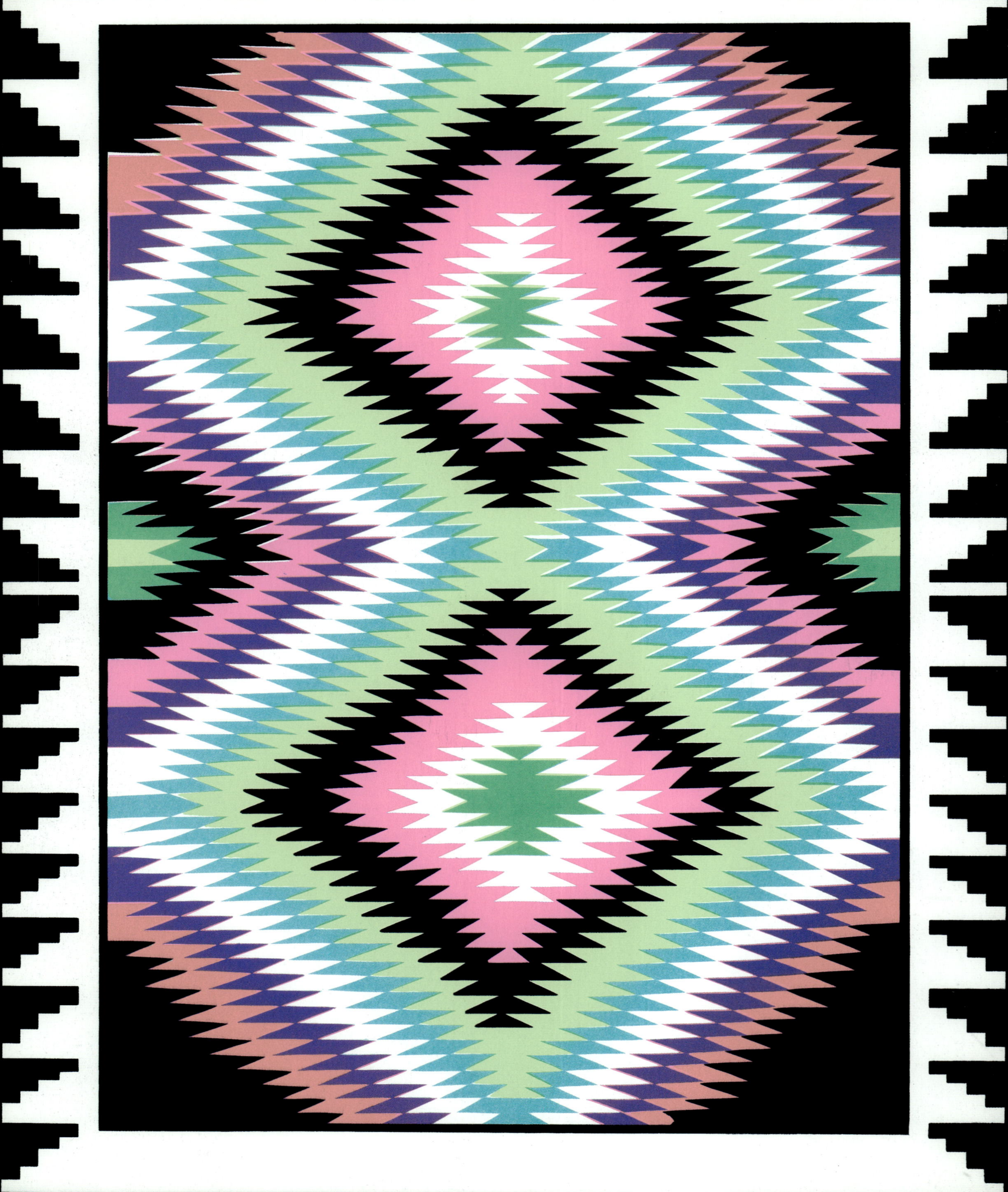

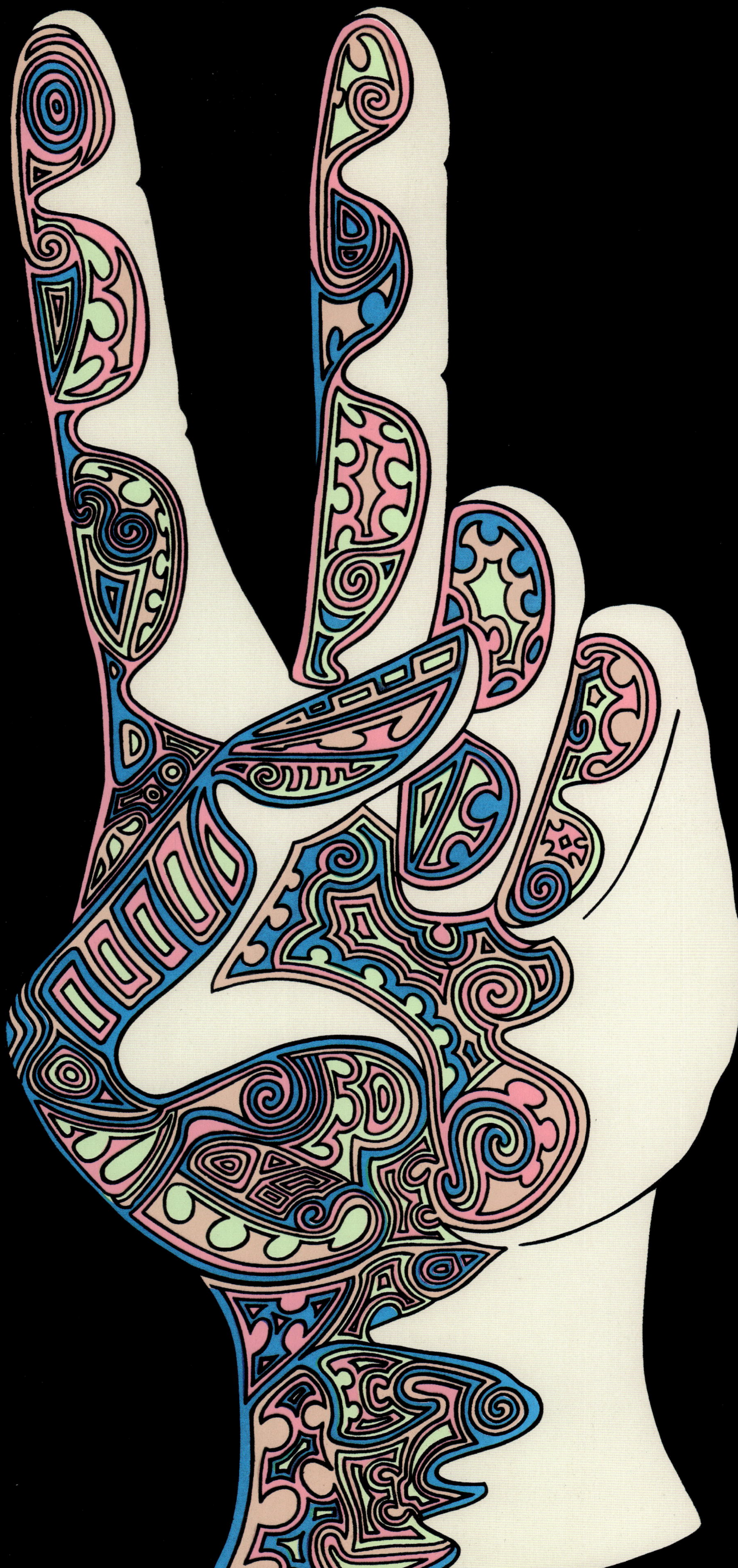

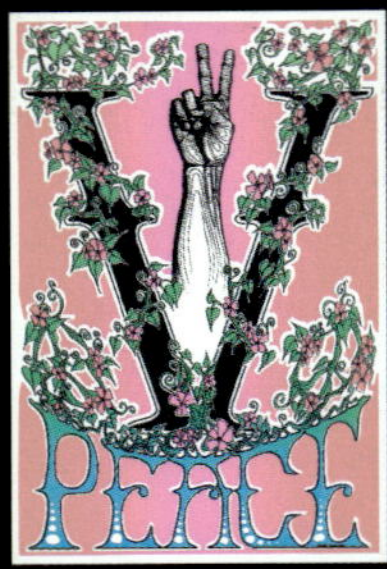

Peace
Matt Boulton

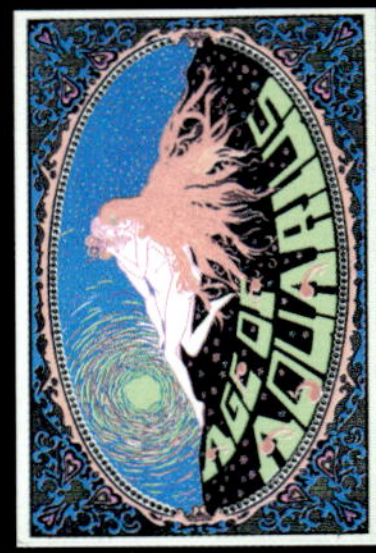

Age of Aquarius

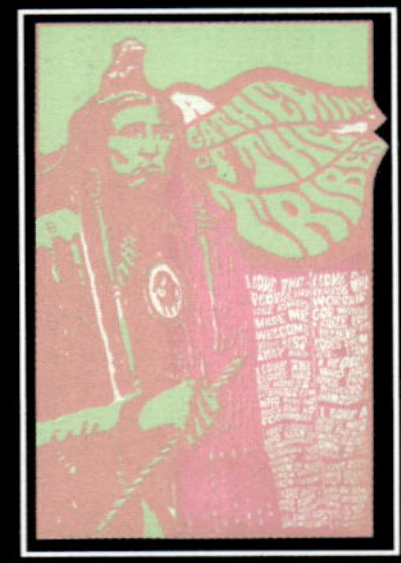

Gathering of the Tribes

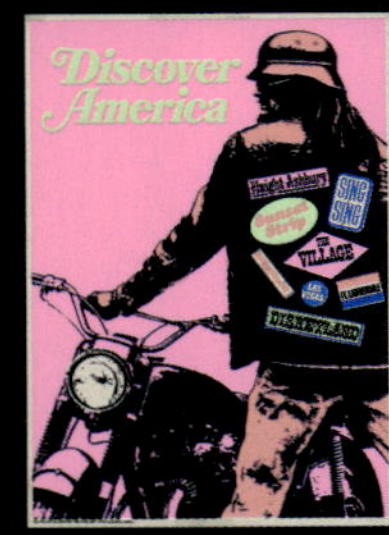

Discover America, 1968
Electric Posters /
Fleer Corp.
Philadelphia, PA

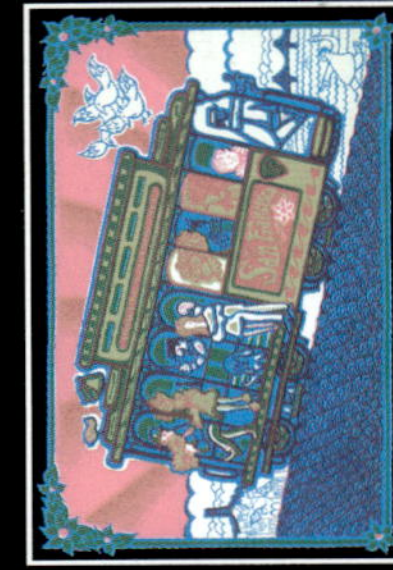

San Francisco
Harry Hambly
Hambly Studios
Santa Clara, CA

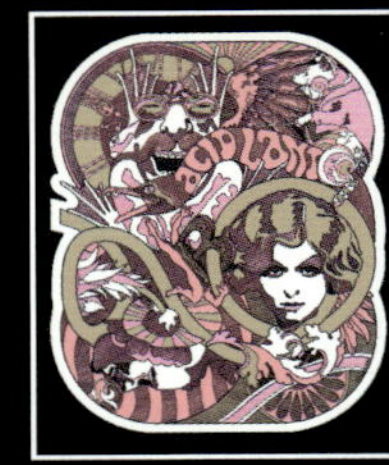

Acid Land, 1967
Accent Posters

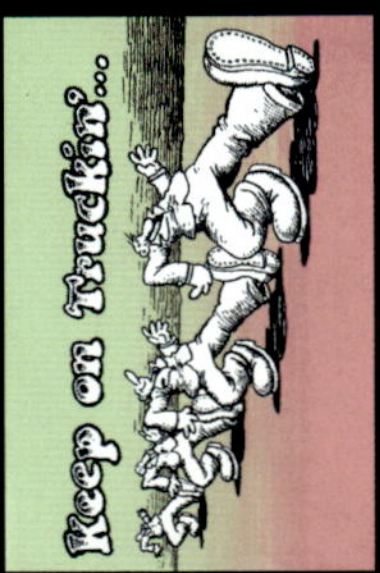

Keep On Truckin'
R. Crumb

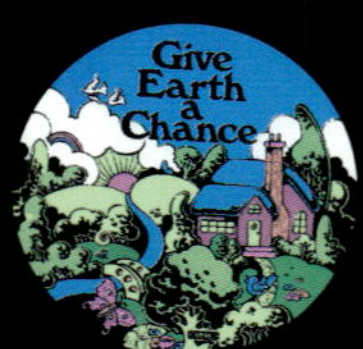

Give Earth a Chance
Ron Rea

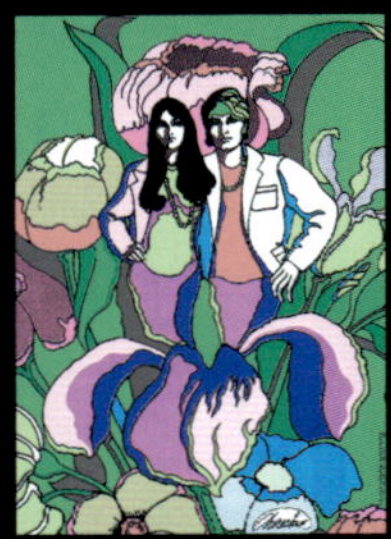

Dear Love
Chereskin
Poster Prints
Conshohocken, PA

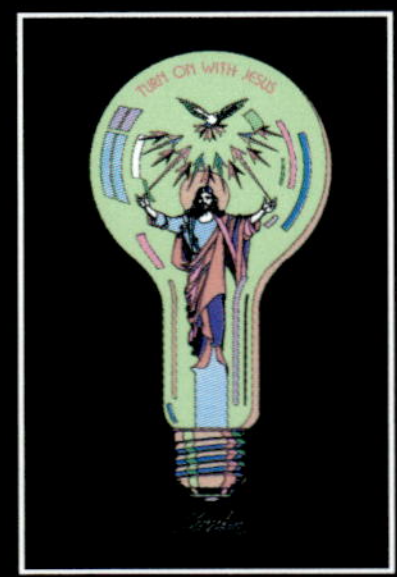

Turn On with Jesus, 1972
Chereskin
Artko Studios
Albuquerque, NM

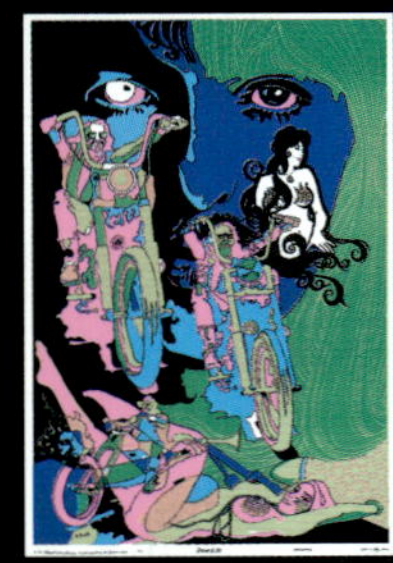

Dream of Me, 1968
B. Moyer
Celestial Arts
Posters
San Francisco, CA

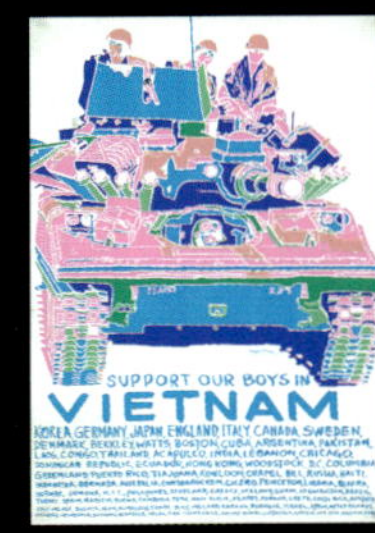

Support Our Boys in Vietnam
Poster Prints
Conshohocken, PA

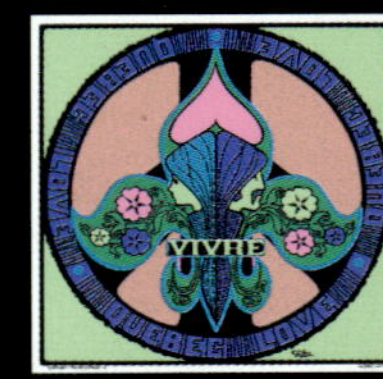

Quebec Love
Giles
Personality Posters
of Canada Ltd.

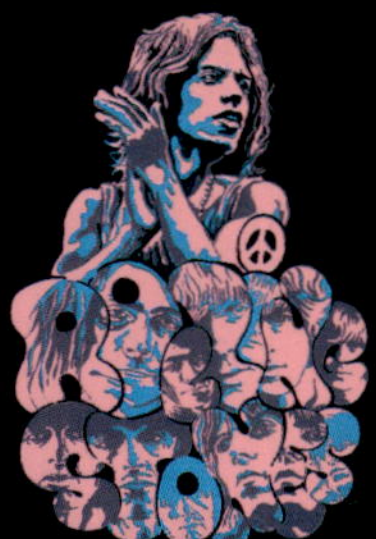

Rolling Stones
Dail W. Beeghly
Creative Signs, Inc.
Silver Spring, MD

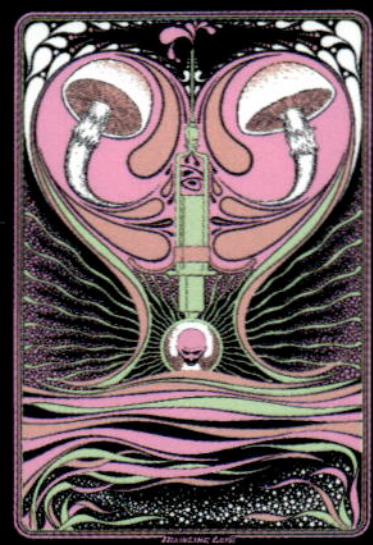

Mainline Love, 1969
Buroff
Frank Kay
Distributors, Inc.
Hollywood, CA

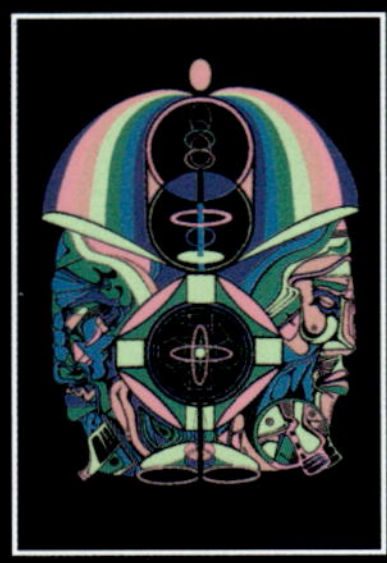

Celestial Odyssey, 1972
Artko Studios
Albuquerque, NM

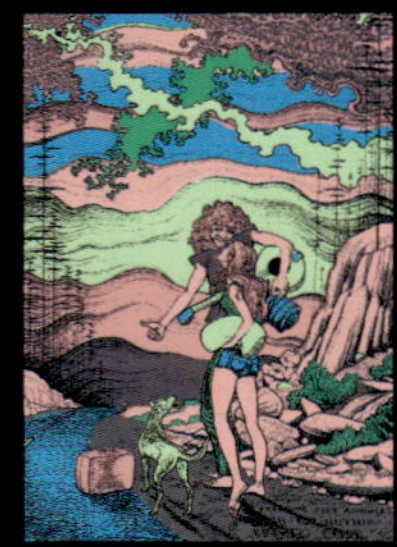

Free Summer, 1971
Goode
Houston Blacklight
& Poster Dist. Co.
Houston, TX

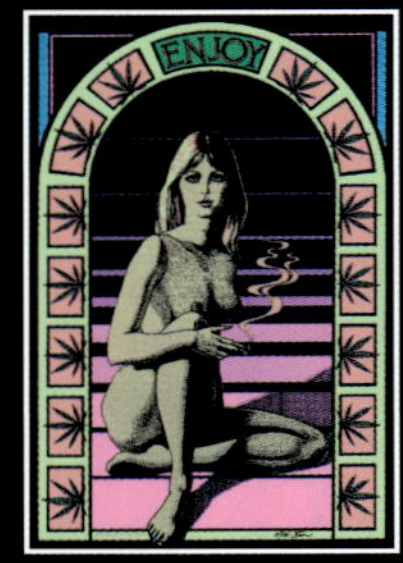

Sunshine, 1976
M.H. Stein
Dynamic Pub Co.
Medina, OH

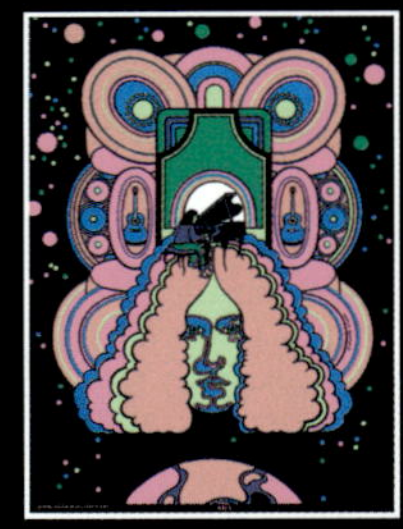

Arlo, 1970
Dream Merchants
New York, NY

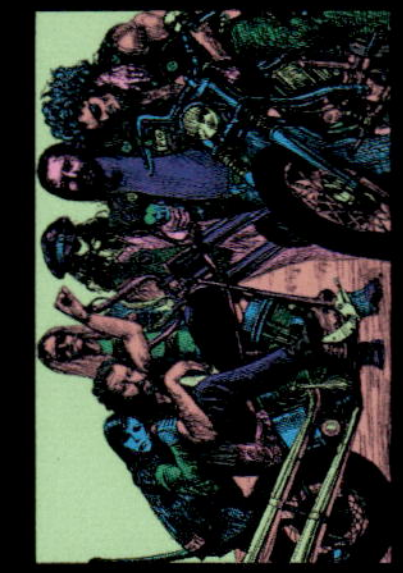

Doin Our Thing, 1969
Houston Blacklight
& Poster Dist. Co.
Houston, TX

Panther
One Stop Posters
Chicago, IL

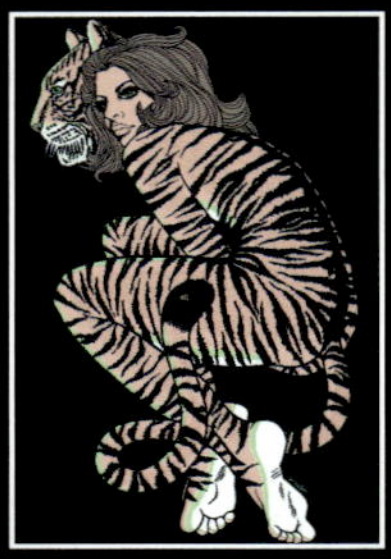

Tiger Lady
Michel Art
Personality Posters
of Canada Ltd.

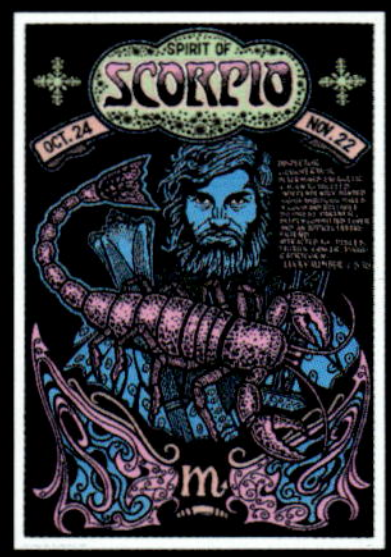

Scorpio
H. Kang Co., Inc.
Mt. Vernon, NY

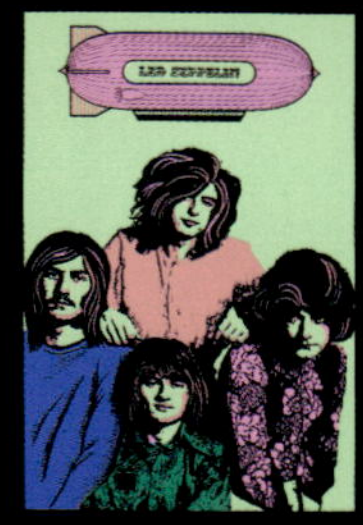

Led Zeppelin

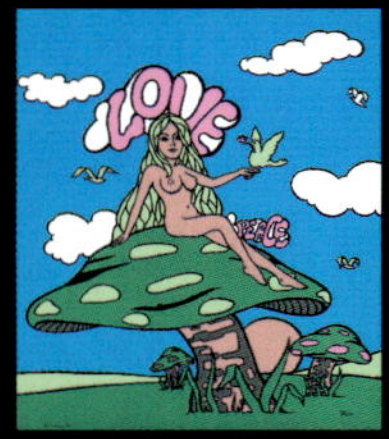

LOVE
D.Wilks
Kaymac

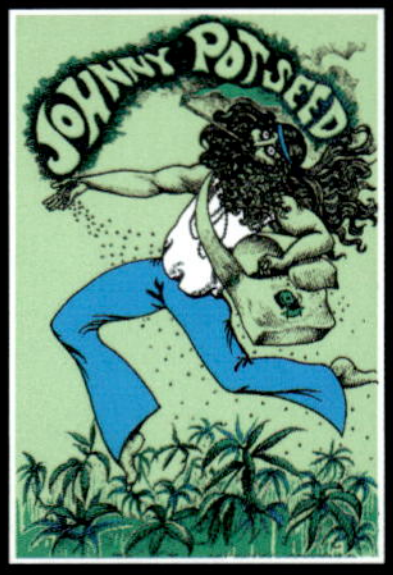

Johnny Potseed, 1969
G.G. Houston
Blacklight & Poster
Dist. Co.
Houston, TX

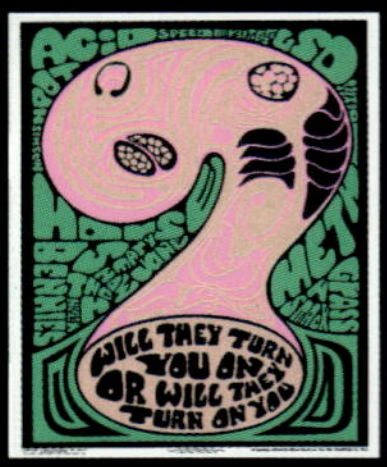

Will They Turn You On..., 1970
The National
Institute of Mental
Health
Washington, D.C.

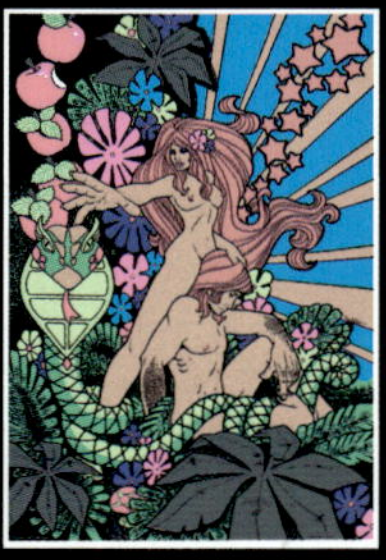

Adam & Even
Michel Art
Personality Posters
of Canada Ltd.

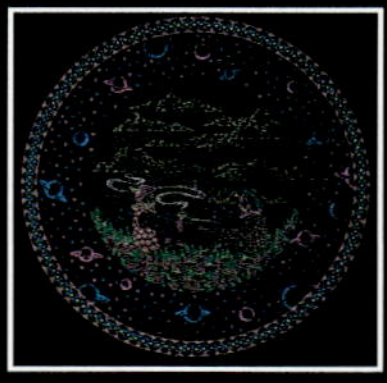

The Eclipse Poster World

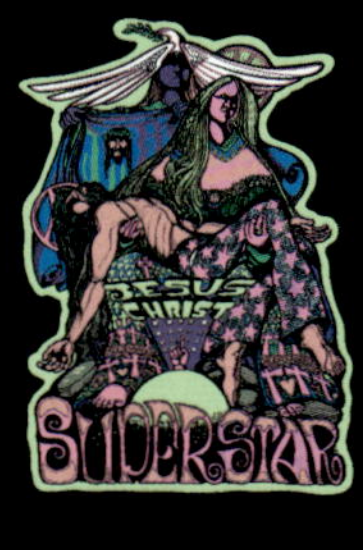

Jesus Christ Superstar II, 1971
Artko-IMS
Albuquerque, NM

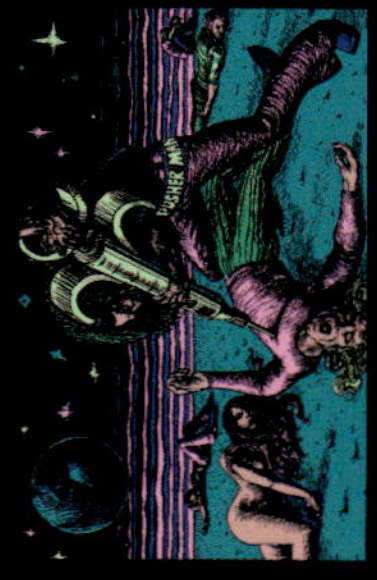

The Losers, 1970
Houston Blacklight
& Poster Dist. Co.
Houston, TX

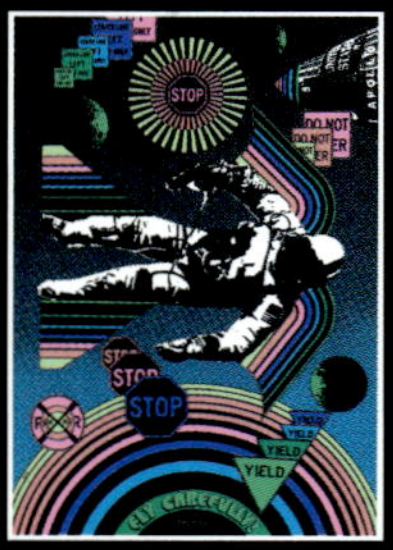

Fly Carefully, 1969
Stanislaw Zagorski
Poster Prints
Conshohocken, PA

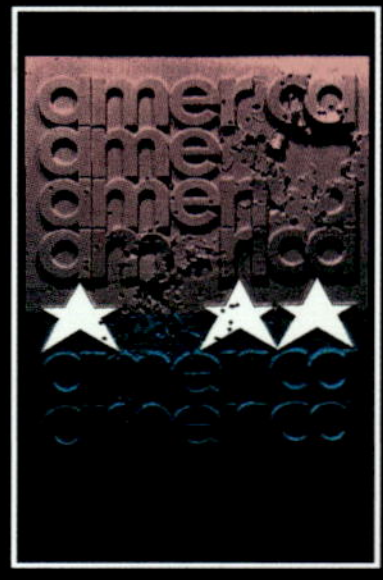

America, 1972
Pro Arts, Inc.
Medina, OH

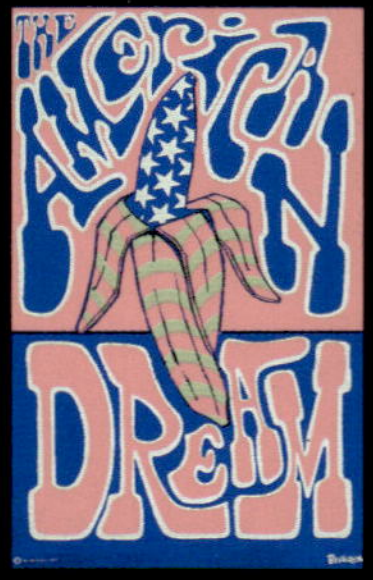

The American Dream, 1967
Art Bevacqua
AA Sales, Inc.
Seattle, WA

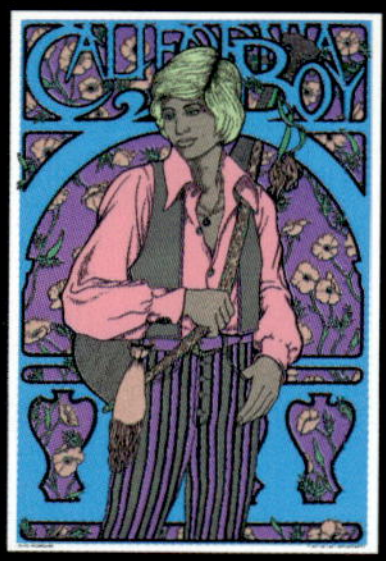

California Boy
AA Sales, Inc.
Seattle, WA

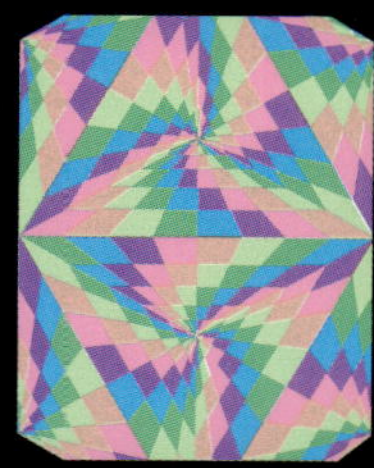

Deep Deep
Nancy Parker
A. Sirkia / Angeles

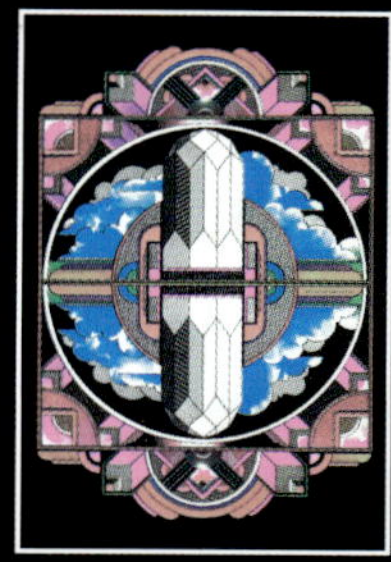

Untitled, 1970
Joe Lombardo
Gemini Rising, Inc.

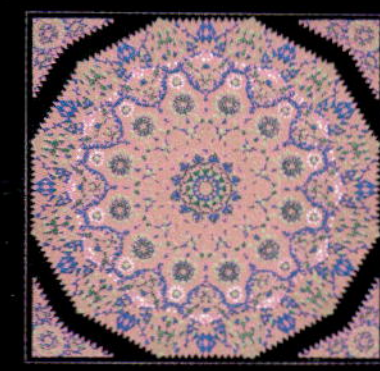

Citadel, 1969
Nancy Parker
Wespac Visual
Communications
San Francisco, CA

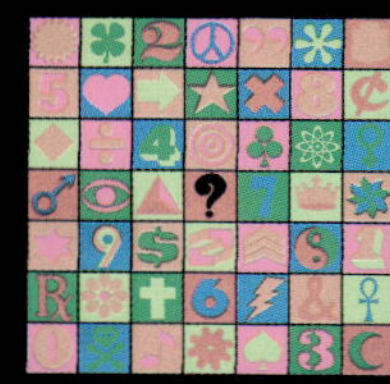

Symbology, 1969
Armand Weston
The Third Eye, Inc.
New York, NY

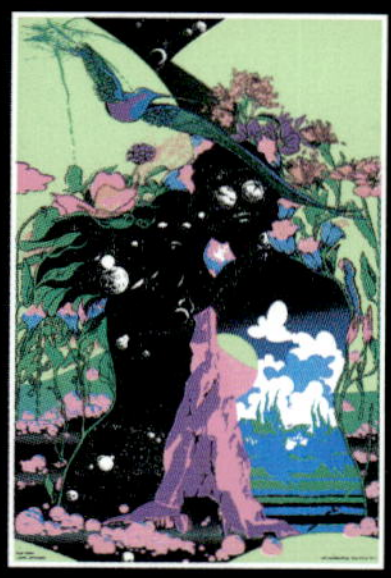

Zephyr, 1970
Jupiter Rubin
The Third Eye, Inc.
New York, NY

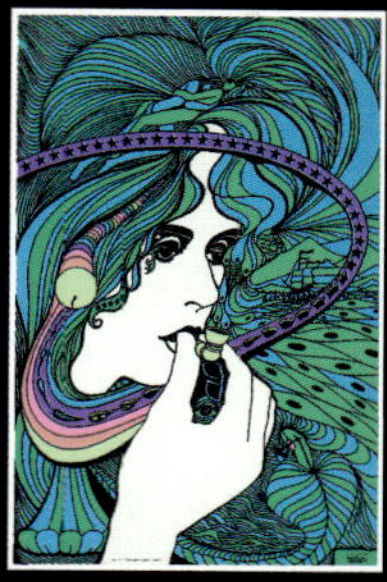

Acid Queen, 1970
Tom Gatz
Pro Arts, Inc.
Medina, OH

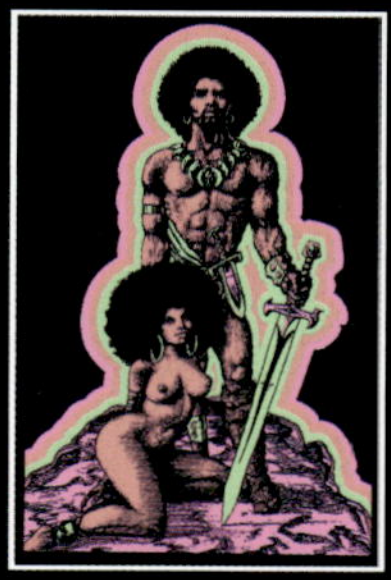

Man and Woman 1,
1970
George F. Goode
Houston Blacklight
& Poster Dist. Co.
Houston, TX

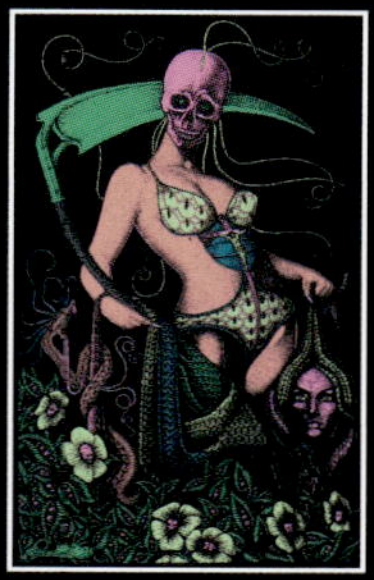

The Devil's Angel,
1971
Broadell
Platt Poster Co.
Los Angeles, CA

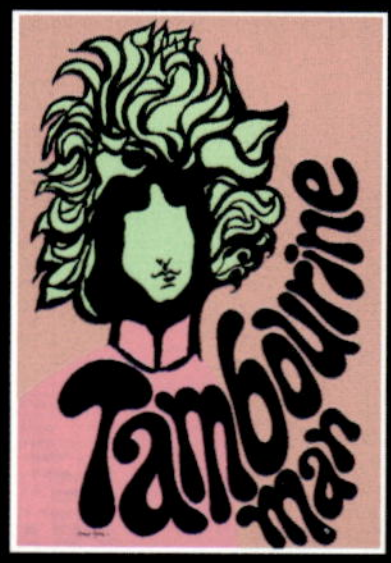

Tambourine Man
Poster Prints
Conshohocken, PA

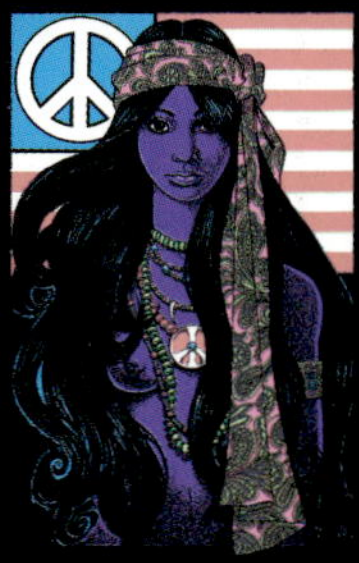

Gypsy Girl
AA Sales, Inc.
Seattle, WA

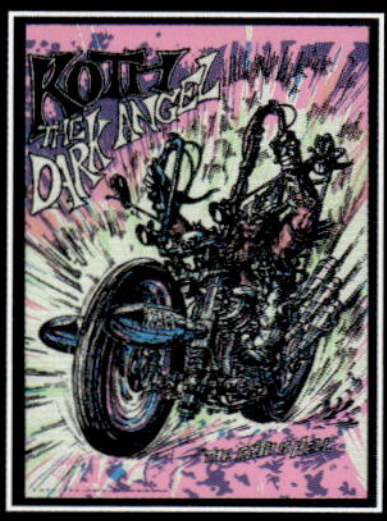

Koth
Western Graphics
Eugene, OR

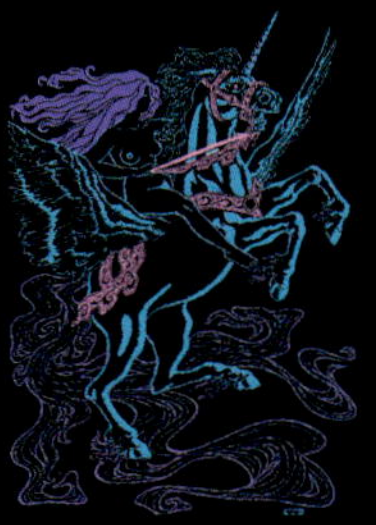

Gypsy Woman, 1971
Ashe Star City, Inc.
Los Angeles, CA

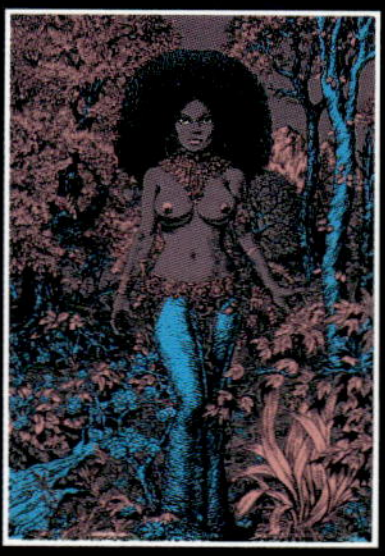

Jungle Princess
George F. Goode
Houston Blacklight
& Poster Dist. Co.
Houston, TX

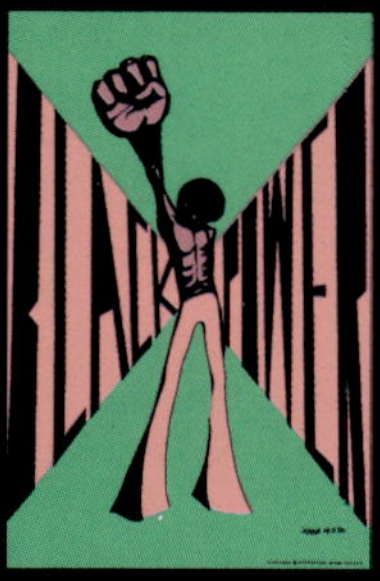

Black Power, 1971
Zash M.J.B.
Pro Arts, Inc.
Medina, OH

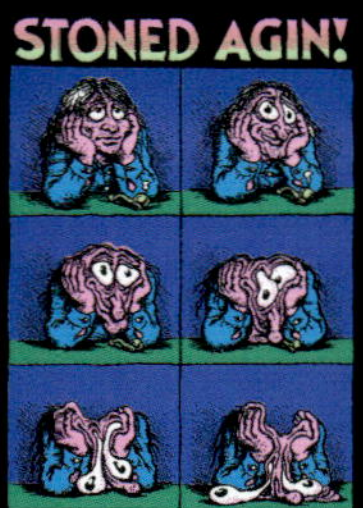

Stoned Again, 1971
R. Crumb
Star City
Distributors, Inc.
Los Angeles, CA

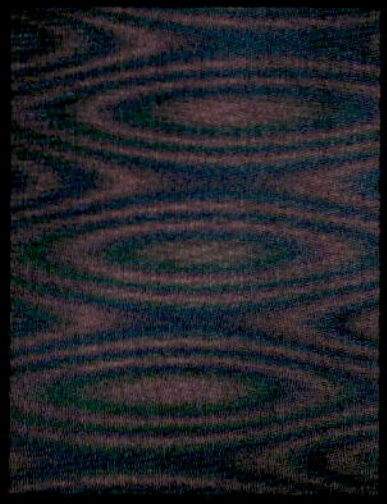

Untitled

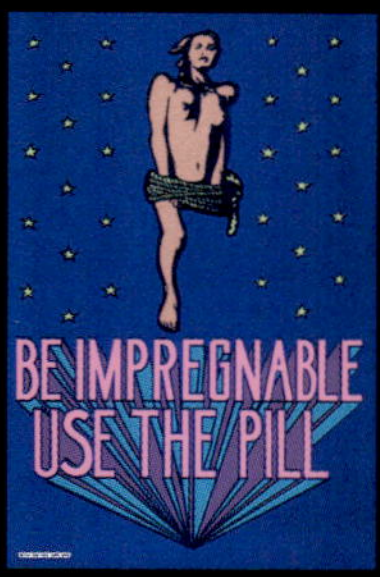

Be Impregnable Use the Pill, 1969
Pro Arts, Inc.
Kent, OH

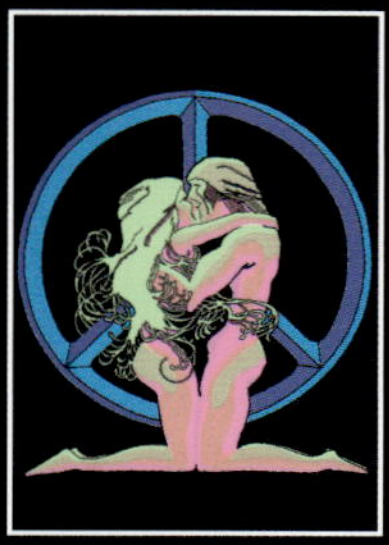

Erotica, 1970
Pro Arts, Inc.
Kent, OH

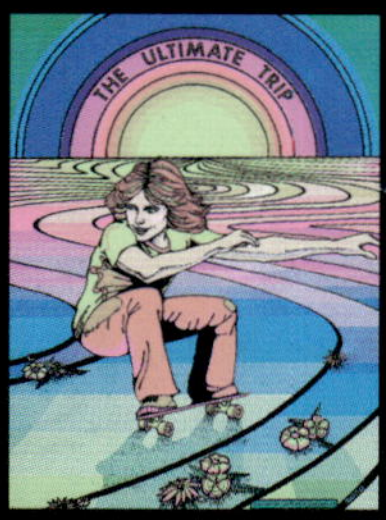

The Ultimate Trip,
1976
Pro Arts, Inc.
Kent, OH

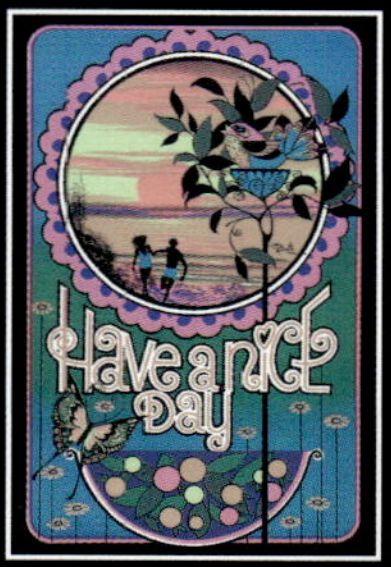

Have a Nice Day,
1970
Star City
Distributors, Inc.
Los Angeles, CA

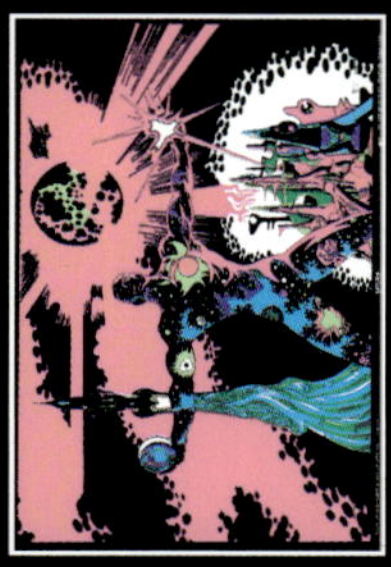

Saturn, 1970
Mitchell
Houston Blacklight
& Poster Dist. Co.
Houston, TX

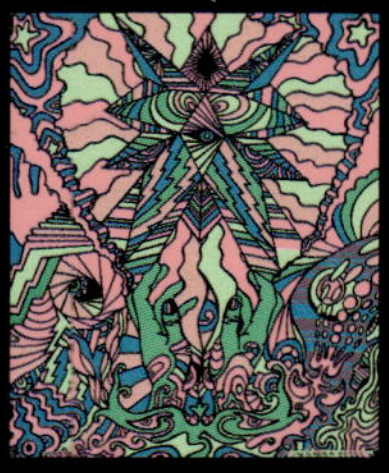

In My Room, 1969
Gary Edwards
The Third Eye, Inc.
New York, NY

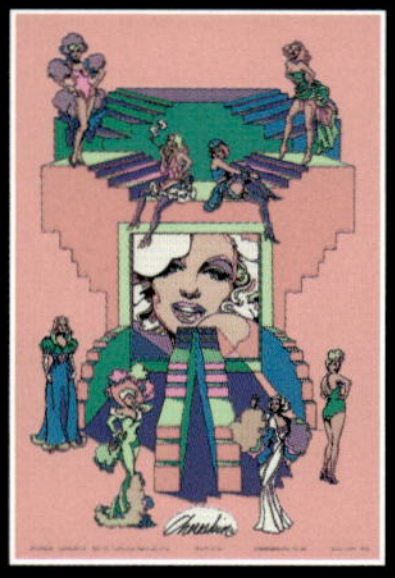

Screen Queens
Chereskin
Artko Studio Arts
Corp.
Albuquerque, NM

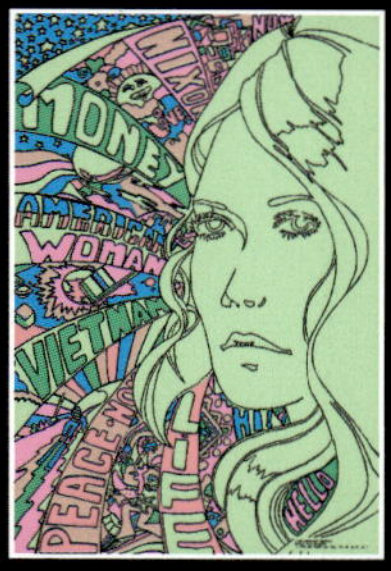

American Woman,
1970
Rick Ambrose
The Third Eye, Inc.
New York, NY

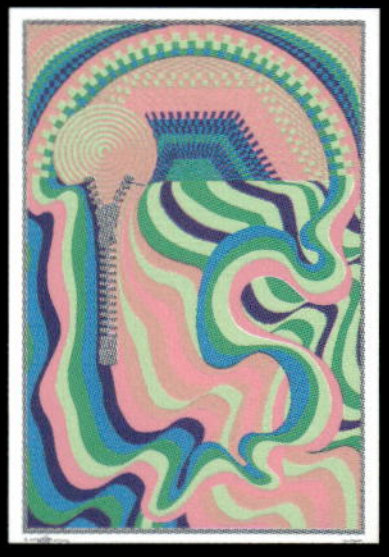

PSZYPP, 1968
Le Roy Olson
Artko Studios

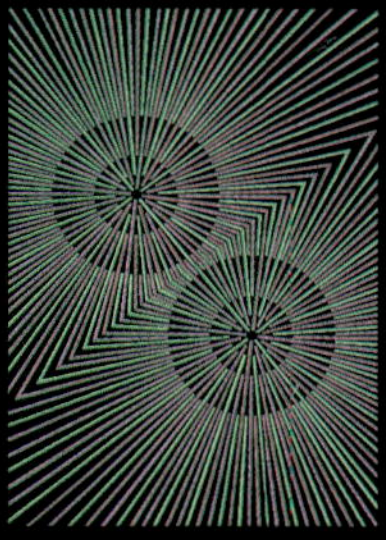

Untitled
Poster Prints
Conshohocken, PA

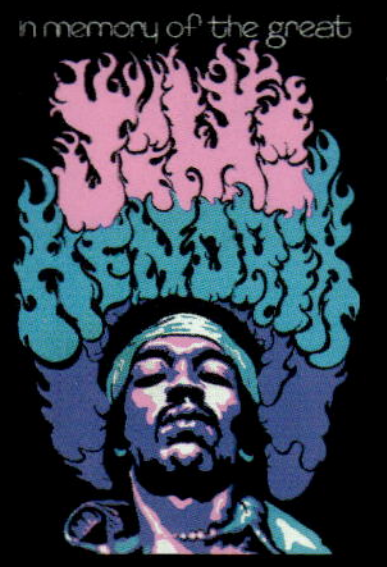

Jimi Hendrix, 1970
Dail W. Beeghly

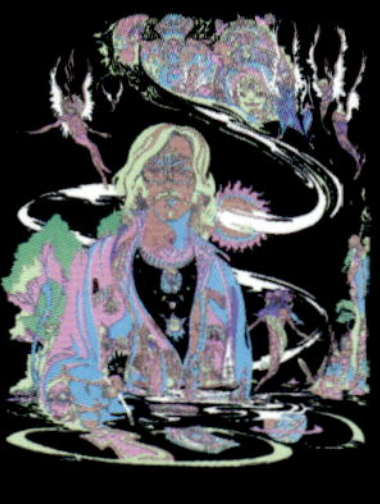

Riding Easy, 1969
Howard Bernstein
Wespac Visual
Communications
San Francisco, CA

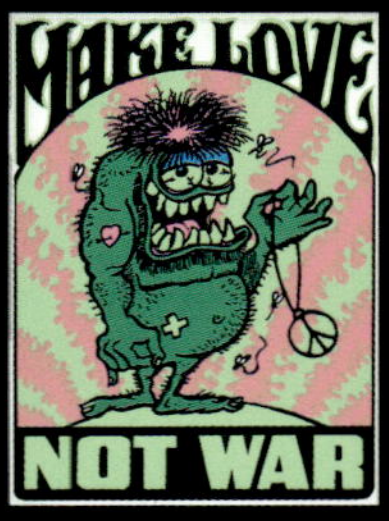

Make Love Not War,
1970
Hip Products
Chicago, IL

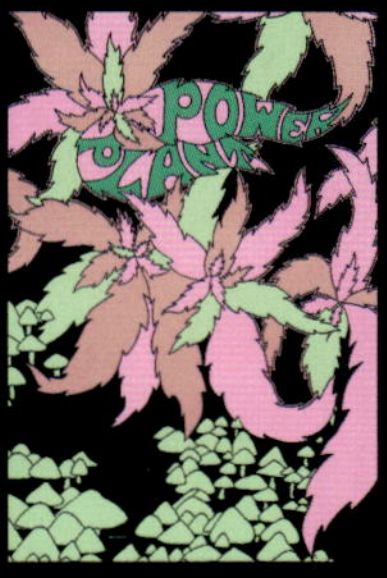

Power Plant, 1968
Orbit Graphic Arts /
Creative Unlimited
San Francisco, CA

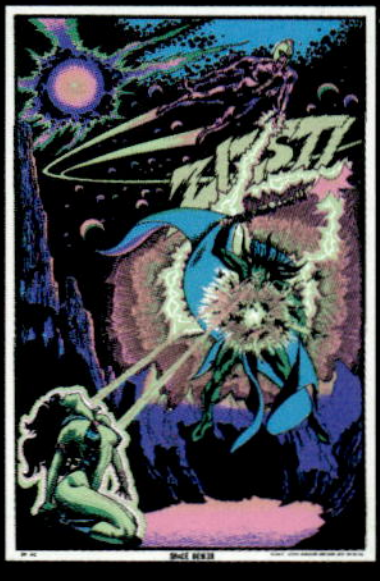

Space Demon, 1972
AA Sales, Inc.
Seattle, WA

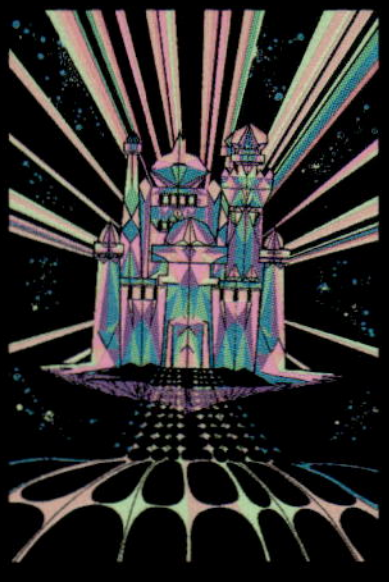

House of Stone, 1969
The Third Eye, Inc.
New York, NY

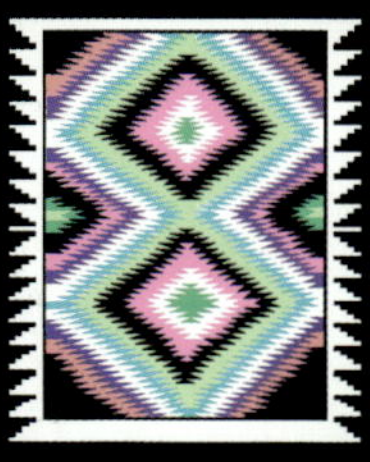

Navajo Op

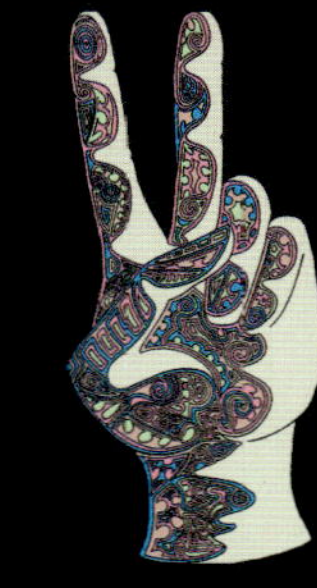

The Hand, 1971
Directional Arts
Distribution

ACKNOWLEDGMENTS

I would like to thank Eva Prinz for getting this project started and David Cashion for seeing it through. I would like to thank Tina Vestin for her tireless work in helping me make this book. I would like to thank all the artists whose work is presented within and Arik Roper and Jeff Sims for keeping the dark arts alive. I would also like to thank MGMT for being MGMT and Tim Bonner for his assistance. Thanks to Lance Bangs, Chris Bilheimer, Loren Cronk, Michael Lachowski, Paul Marlow, Noel Tolentino, and Rob and Leslie Vonderheide for their advice in life, and a very special thanks to James Huggins for his constant encouragement. I owe my family a lifetime of thank yous and this is a great chance. Thank you.

—Dan Donahue

The publisher would like to acknowledge the kind help of Blacklight.com.

Blacklight.com, a leading online superstore, offers thousands of lighting, celebration, and security items with an extensive ultraviolet, blacklight reactive, and photoluminescent product selection. Blacklight is our passion, service is our specialty.